Lecture Notes in Computer Science   11038

Commenced Publication in 1973
Founding and Former Series Editors:
Gerhard Goos, Juris Hartmanis, and Jan van Leeuwen

More information about this series at http://www.springer.com/series/7412

Danail Stoyanov · Zeike Taylor
Seyed Mostafa Kia · Ipek Oguz
Mauricio Reyes et al. (Eds.)

Understanding and Interpreting Machine Learning in Medical Image Computing Applications

First International Workshops
MLCN 2018, DLF 2018, and iMIMIC 2018
Held in Conjunction with MICCAI 2018
Granada, Spain, September 16-20, 2018
Proceedings

 Springer

Editors
Danail Stoyanov
University College London
London, UK

Zeike Taylor
University of Leeds
Leeds, UK

Seyed Mostafa Kia ⓘ
Radboud University Medical Center
Nijmegen, The Netherlands

Ipek Oguz ⓘ
Vanderbilt University
Nashville, TN, USA

Mauricio Reyes ⓘ
University of Bern
Bern, Switzerland

Additional Workshop Editors *see next page*

ISSN 0302-9743   ISSN 1611-3349   (electronic)
Lecture Notes in Computer Science
ISBN 978-3-030-02627-1   ISBN 978-3-030-02628-8   (eBook)
https://doi.org/10.1007/978-3-030-02628-8

Library of Congress Control Number: 2018958518

LNCS Sublibrary: SL6 – Image Processing, Computer Vision, Pattern Recognition, and Graphics

This Springer imprint is published by the registered company Springer Nature Switzerland AG
The registered company address is: Gewerbestrasse 11, 6330 Cham, Switzerland

Additional Workshop Editors

Tutorial and Educational Chair

Anne Martel
University of Toronto
Toronto, ON
Canada

Workshop and Challenge Co-chair

Lena Maier-Hein
German Cancer Research Center (DKFZ)
Heidelberg
Germany

First International Workshop on Machine Learning in Clinical Neuroimaging, MLCN 2018

Andre F. Marquand
Radboud University
Nijmegen
The Netherlands

Edouard Duchesnay ⓘ
NeuroSpin, CEA
Saclay, Paris
France

Tommy Löfstedt ⓘ
Umeå University
Umeå
Sweden

First International Workshop on Deep Learning Fails, DLF 2018

Bennett Landman ⓘ
Vanderbilt University
Nashville, TN
USA

M. Jorge Cardoso ⓘ
King's College London
London
UK

First International Workshop on Interpretability of Machine Intelligence in Medical Image Computing, iMIMIC 2018

Carlos A. Silva ⓘ
University of Minho
Guimarães
Portugal

Sérgio Pereira ⓘ
University of Minho
Guimarães
Portugal

Raphael Meier
University of Bern
Bern
Switzerland

MLCN 2018 Preface

The first international workshop on Machine Learning in Clinical Neuroimaging (MLCN) was held in conjunction with MICCAI 2018, with a special focus on spatially structured data analysis. The workshop aimed to bring together top-notch researchers in machine learning and clinical neuroscience to discuss and hopefully bridge the existing gap in applied machine learning in clinical neuroscience. The main objective was to shed light on the opportunities and challenges in the structure-aware modeling in neuroimaging data in both encoding and decoding settings. For the keynote talks, leading researchers in the domain of spatial statistics, pattern recognition in neuroimaging, and predictive clinical neuroscience, Prof. Christos Davatzikos (University of Pennsylvania), Dr. Gael Varoquaux (Inria), and Dr. George Langs (Medical University of Vienna), were invited in order to provide a comprehensive overview from theory to application in the field.

The call for papers for the MLCN 2018 workshop was released on April 1, 2018, with the paper deadline set to July 25, 2018. The seven manuscripts received went through a double-blind review process by the MLCN scientific committee (Ehsan Adeli, Andre Altman, Luca Ambrogioni, Richard Dinga, Koen Haak, Christina Isakoglou, Emanuele Olivetti, Pradeep Reddy Raamana, Kerstin Ritter, Sourena Soheili Nezhad, Thomas Wolfers, and Maryam Zabihi), and the top four papers with the best reviews were accepted for publication in the proceedings. The accepted contributions develop state-of-the-art machine learning methods such as spatio-temporal Gaussian process analysis, stochastic variational inference, and deep learning for applications in Alzheimer's disease diagnosis and multi-site neuroimaging data analysis.

September 2018

Seyed Mostafa Kia
Andre Marquand
Edouard Duchesnay
Tommy Löfstedt

DLF 2018 Preface

Deep learning methods have rapidly become omnipresent within the medical image computing and computer assisted intervention (MICCAI) community in recent years, thanks to their many attractive properties including state-of-the-art accuracy in many tasks in areas such as segmentation and classification. However, now that the initial excitement about these new techniques has led to many successful applications within the MICCAI domain, we need to begin developing a better understanding to demystify deep learning.

To this end, we hosted a new workshop in conjunction with MICCAI 2018 (September 16–20, 2018, Granada, Spain) dedicated to understanding the "edges" of deep learning: What are its current limitations? What are some MICCAI problems that are not well-suited for existing DL methods? What are some failures the community has encountered in DL? How can we better understand the "mysteries" we encounter, whether an algorithm works unexpectedly well or unexpectedly poorly? Where is the field going? etc. The workshop was held on September 16, 2018, in the Granada Exhibition and Conference Center.

Submissions were solicited via a call for papers that was widely circulated. Some example ideas for possible contributions were suggested, but more importantly, we invited the MICCAI community to brainstorm about deep learning in the context of MICCAI-related fields. We were pleased to observe the community responded well to this invitation and the submissions covered a wide range of topics. Each submission underwent a single-blind review by at least two members of the Program Committee, which consisted of researchers who actively contribute in the area. At the conclusion of the review process, six papers were accepted. The program was further enriched by two invited speakers, Dr. Scott Acton and Dr. Leo Grady.

We chose the (provocative on purpose) title of "Deep Learning Fails" for this workshop, tongue-in-cheek. To be clear, the goal of this workshop was *not* to disparage deep learning methods. As such, we did not allow papers that indiscriminately trivialize deep learning, such as a paper showing negative results using a generic network model that has not been adapted or fine-tuned to address a specific problem. Rather, we encouraged submissions that are in the spirit of constructive criticism, with the aim of evaluating the strengths and weaknesses of deep learning, as well as identifying the main challenges in the current state-of-the-art and future directions.

We would like to thank everyone who helped make this workshop happen: the authors who contributed their work, the Program Committee for their careful and thoughtful review, the invited speakers for sharing their expertise and insights, the attendees for their contribution to the discussion, and the MICCAI society for general support.

September 2018

Ipek Oguz
Bennett Landman
Jorge M. Cardoso

iMIMIC 2018 Preface

The first edition of the workshop on Interpretability of Machine Intelligence in Medical Image Computing (iMIMIC)[1] was held on September 16, 2018, as a half-day satellite event of the 21st International Conference on Medical Image Computing and Computer-Assisted Intervention (MICCAI), in Granada, Spain. With its first edition, this workshop aimed at introducing the challenges and opportunities of interpretability of machine learning systems in the context of MICCAI, as well as understanding the current state of the art in the topic and promoting it as a crucial area for further research. The workshop program comprised of oral presentations of the accepted works and two keynotes provided by experts in the field.

Machine learning systems are achieving remarkable performances at the cost of increased complexity. Hence, they have become less interpretable, which may cause distrust. As these systems are pervasively being introduced to critical domains, such as medical image computing and computer-assisted intervention, it becomes imperative to develop methodologies to explain their predictions. Such methodologies would help physicians to decide whether they should follow/trust a prediction. Additionally, it could facilitate the deployment of such systems, from a legal perspective. Ultimately, interpretability is closely related to AI safety in health care. Besides increasing trust and acceptance by physicians, interpretability of machine learning systems can be helpful for other purposes, such as during method development, for revealing biases in the training data, or studying and identifying the most relevant data (e.g., specific MRI sequences in multi-sequence acquisitions).

The iMIMIC proceedings include six 8-page papers carefully selected from a larger pool of submitted manuscripts, following a rigorous single-blinded peer-review process. Each paper was reviewed by at least two expert reviewers. All the accepted papers were presented as oral presentations during the workshop, with time for questions and discussion.

We thank all the authors for their participation and the Program Committee members for contributing to this workshop. We are also very grateful to our sponsors H2O. ai, the Competence Center for Medical Technology (CCMT), Olea Medical, and the support of the promotion fund for young researchers at the University of Bern.

September 2018

Mauricio Reyes
Carlos A. Silva
Sérgio Pereira
Raphael Meier

---

[1] https://imimic.bibbucket.io/.

Organization

MLCN 2018 Organizing Committee

Andre Marquand   Radbourg University, The Netherlands
Edouard Duchesnay   CEA NeuroSpin, France
Tommy Löfstedt   Umeå University, Sweden
Seyed Mostafa Kia   Radboud University, The Netherlands

DLF 2018 Organizing Committee

Ipek Oguz   University of Pennsylvania, USA
Bennett Landman   Vanderbilt University, USA
Jorge Cardoso   King's College London, UK

iMIMIC 2018 Organizing Committee

Mauricio Reyes   University of Bern, Switzerland
Carlos A. Silva   University of Minho, Portugal
Sérgio Pereira   University of Minho, Portugal
Raphael Meier   University of Bern, Switzerland

Contents

First International Workshop on Machine Learning in Clinical Neuroimaging, MLCN 2018

Alzheimer's Disease Modelling and Staging Through Independent
Gaussian Process Analysis of Spatio-Temporal Brain Changes 3
    Clement Abi Nader, Nicholas Ayache, Philippe Robert,
    Marco Lorenzi, and for the Alzheimer's Disease Neuroimaging Initiative

Multi-channel Stochastic Variational Inference for the Joint Analysis of
Heterogeneous Biomedical Data in Alzheimer's Disease 15
    Luigi Antelmi, Nicholas Ayache, Philippe Robert,
    Marco Lorenzi, and for the Alzheimer's Disease Neuroimaging Initiative

Visualizing Convolutional Networks for MRI-Based Diagnosis of
Alzheimer's Disease 24
    Johannes Rieke, Fabian Eitel, Martin Weygandt, John-Dylan Haynes,
    and Kerstin Ritter

Finding Effective Ways to (Machine) Learn fMRI-Based Classifiers
from Multi-site Data 32
    Roberto Vega and Russ Greiner

First International Workshop on Deep Learning Fails Workshop, DLF 2018

Towards Robust CT-Ultrasound Registration Using Deep
Learning Methods 43
    Yuanyuan Sun, Adriaan Moelker, Wiro J. Niessen,
    and Theo van Walsum

To Learn or Not to Learn Features for Deformable Registration? 52
    Aabhas Majumdar, Raghav Mehta, and Jayanthi Sivaswamy

Evaluation of Strategies for PET Motion Correction - Manifold Learning
vs. Deep Learning 61
    James R. Clough, Daniel R. Balfour, Claudia Prieto, Andrew J. Reader,
    Paul K. Marsden, and Andrew P. King

Exploring Adversarial Examples: Patterns of One-Pixel Attacks 70
    David Kügler, Alexander Distergoft, Arjan Kuijper,
    and Anirban Mukhopadhyay

Shortcomings of Ventricle Segmentation Using Deep
Convolutional Networks.   79
    Muhan Shao, Shuo Han, Aaron Carass, Xiang Li, Ari M. Blitz,
    Jerry L. Prince, and Lotta M. Ellingsen

Vulnerability Analysis of Chest X-Ray Image Classification Against
Adversarial Attacks.   87
    Saeid Asgari Taghanaki, Arkadeep Das, and Ghassan Hamarneh

**First International Workshop on Interpretability of Machine Intelligence
in Medical Image Computing, iMIMIC 2018**

Collaborative Human-AI (CHAI): Evidence-Based Interpretable Melanoma
Classification in Dermoscopic Images   97
    Noel C. F. Codella, Chung-Ching Lin, Allan Halpern, Michael Hind,
    Rogerio Feris, and John R. Smith

Automatic Brain Tumor Grading from MRI Data Using Convolutional
Neural Networks and Quality Assessment   106
    Sérgio Pereira, Raphael Meier, Victor Alves, Mauricio Reyes,
    and Carlos A. Silva

Visualizing Convolutional Neural Networks to Improve Decision Support
for Skin Lesion Classification.   115
    Pieter Van Molle, Miguel De Strooper, Tim Verbelen,
    Bert Vankeirsbilck, Pieter Simoens, and Bart Dhoedt

Regression Concept Vectors for Bidirectional Explanations
in Histopathology   124
    Mara Graziani, Vincent Andrearczyk, and Henning Müller

Towards Complementary Explanations Using Deep Neural Networks   133
    Wilson Silva, Kelwin Fernandes, Maria J. Cardoso,
    and Jaime S. Cardoso

How Users Perceive Content-Based Image Retrieval for Identifying
Skin Images.   141
    Mahya Sadeghi, Parmit K. Chilana, and M. Stella Atkins

Author Index   149

First International Workshop on Machine Learning in Clinical Neuroimaging, MLCN 2018

Alzheimer's Disease Modelling and Staging Through Independent Gaussian Process Analysis of Spatio-Temporal Brain Changes

Clement Abi Nader[1]([✉]), Nicholas Ayache[1], Philippe Robert[2,3],
Marco Lorenzi[1], and for the Alzheimer's Disease Neuroimaging Initiative

[1] UCA, Inria Sophia Antipolis, Epione Research Project, Sophia Antipolis, France
{clement.abi-nader,nicholas.ayache,marco.lorenzi}@inria.fr
[2] UCA, CoBTeK, Nice, France
[3] Centre Memoire, CHU de Nice, Nice, France
probert@unice.fr

Abstract. Alzheimer's disease (AD) is characterized by complex and largely unknown progression dynamics affecting the brain's morphology. Although the disease evolution spans decades, to date we cannot rely on long-term data to model the pathological progression, since most of the available measures are on a short-term scale. It is therefore difficult to understand and quantify the temporal progression patterns affecting the brain regions across the AD evolution. In this work, we present a generative model based on probabilistic matrix factorization across temporal and spatial sources. The proposed method addresses the problem of disease progression modelling by introducing clinically-inspired statistical priors. To promote smoothness in time and model plausible pathological evolutions, the temporal sources are defined as monotonic and independent Gaussian Processes. We also estimate an individual time-shift parameter for each patient to automatically position him/her along the sources time-axis. To encode the spatial continuity of the brain substructures, the spatial sources are modeled as Gaussian random fields. We test our algorithm on grey matter maps extracted from brain structural images. The experiments highlight differential temporal progression patterns mapping brain regions key to the AD pathology, and reveal a disease-specific time scale associated with the decline of volumetric biomarkers across clinical stages.

Data used in preparation of this article were obtained from the Alzheimer's Disease Neuroimaging Initiative (ADNI) database (adni.loni.usc.edu). As such, the investigators within the ADNI contributed to the design and implementation of ADNI and/or provided data but did not participate in analysis or writing of this report. A complete listing of ADNI investigators can be found at: http://adni.loni.usc.edu/wp-content/uploads/how_to_apply/ADNI_Acknowledgement_List.pdf.

D. Stoyanov et al. (Eds.): MLCN 2018/DLF 2018/iMIMIC 2018, LNCS 11038, pp. 3–14, 2018.
https://doi.org/10.1007/978-3-030-02628-8_1

1   Introduction

Neurodegenerative disorders such as Alzheimer's disease (AD) are characterized by morphological and molecular changes of the brain, and ultimately lead to cognitive and behavioral decline [8]. To date there is no clear understanding of the dynamics regulating the disease progression. Consequently several attempts have been made to model the disease evolution in a data-driven way, using sets of biomarkers extracted from different imaging acquisition techniques, such as Magnetic Resonance Imaging (MRI) [12]. However available data are mostly represented by cross-sectional measures or time-series acquired on a short-term time span, while the ultimate goal is to unveil the "long-term" disease evolution spreading over decades. Therefore there is a critical need to define the AD evolution in a data-driven manner with respect to an absolute time scale associated to the natural history of the pathology.

To this end, in [9] the authors introduce a disease progression score for each patient in order to identify a data-driven disease scale. This score is based on a set of biomarkers and was shown to correlate with the decline of brain cognitive abilities. A similar approach was proposed by [12] and [6] with scalar biomarkers. In [3], a disease progression score was estimated using higher-dimensional biomarkers from molecular imaging. However these methods don't provide information about the brain structures involved in AD, and how the disease affects them along time. To overcome these limitations, [13] proposes a spatio-temporal model of disease progression explicitly accounting for different temporal dynamics across the brain. This is done by decomposing cortical thickness measurements as a mixture of spatio-temporal processes, by associating each vertex to a temporal progression modeled by a sigmoid function. They also estimate a disease progression score for each subject as a linear transformation of time. However since the proposed formulation does not account for spatial correlation between vertices, it may be potentially sensitive to spatial variation and noise, thus leading to poor interpretability.

The challenge of spatio-temporal modelling in brain images is a classical problem widely addressed via Independent Component Analysis (ICA [7]), especially on functional MRI (fMRI) data [4]. ICA aims at decomposing the data via matrix factorization, looking for a reduced number of spatio-temporal latent sources. Although successful in fMRI analysis, ICA cannot find straightforward applications to the modelling of AD progression. First, ICA retrieves maximally independent latent sources best explaining the data. However, although brain regions can exhibit different atrophy rates, this doesn't necessarily imply statistical independence between them. Second, differently from fMRI data, the absolute time axis of AD spatio-temporal observations is unknown. Thus estimating the pathology timing is a key step in order to model the disease progression, and cannot be performed with standard dimensionality reduction methods such as ICA. Finally, fMRI time series are defined over hundreds of time points, while we work essentially in a cross-sectional setting with one or a few images per-subject.

In this work we present a novel spatio-temporal generative model of disease progression aimed at quantifying the independent dynamics of changes in the brain. We model the observed data through matrix factorization across temporal and spatial sources, with a plausibility constraint introduced by clinically-inspired statistical priors. To promote smoothness in time and model steady evolution from normal to pathological stages, the temporal sources are defined as monotonic independent Gaussian Processes (GPs). We also estimate an individual time-shift parameter for each patient to automatically position him along the sources time-axis. To encode the spatial continuity of the brain sub-structures, the spatial sources are modeled as Gaussian random fields. The framework is efficiently optimized through stochastic variational inference. In the next sections we detail the method formulation and show its application on synthetic and real data composed by a large dataset of MRIs from the Alzheimer's Disease Neuroimaging Initiative (ADNI). Further information can be found in the Appendix.[1]

2   Method

We assume that the spatio-temporal data $Y(x,t) = [Y_1(x,t_1), Y_2(x,t_2), .., Y_P(x,t_p)]$ is stored in a matrix with dimensions $P \times F$, where P is the number of patients, F the number of image features, and $Y_i(x,t_i)$ is the image of an individual i observed at position x and at time t_i. We postulate a generative model in order to decompose the data in N_s spatio-temporal sources such that:

$$Y_p(x, t_p) = S(\theta, t + t_p)A(\psi, x) + \mathcal{E} \tag{1}$$

where S is a $P \times N_s$ matrix where each column represents a temporal trajectory, t_p the individual time-shift parameter, and θ the set of parameters related to the temporal sources. A is a $N_s \times F$ matrix where each row represents a spatial map, and ψ is a set of spatial parameters. $\mathcal{E}$ is a $\mathcal{N}(0, \sigma^2 I)$ Gaussian noise. According to the generative model the likelihood is:

$$p(Y|A, S, \sigma) = \prod_{p=1}^{P} \frac{1}{(2\pi\sigma^2)^{\frac{F}{2}}} \exp(-\frac{1}{2\sigma^2}||Y_p - S(\theta, t + t_p)A(\psi, x)||^2) \tag{2}$$

For each row A_n of A we specify a $\mathcal{N}(0, I)$ prior, while each column S_n of S is a GP modeled as in [5]. This setting leverages on kernel approximation through sampling of basis functions in the spectral domain [14]. For specific choices of the covariance, such as the Radial Basis Function used in our work, the GPs can be approximated as a Bayesian neural network with form: $S(t) = \phi(\Omega t)W$. Where Ω is the projection in the spectral domain, ϕ the non-linear basis function activation, and W the regression parameter. The GPs inference problem thus amounts at estimating approximated distributions for Ω and W.

To account for the steady increase of the sources from normal to pathological stages we introduce a monotonicity prior over the GPs. To do so, we constrain the space of the temporal sources to the set $\mathcal{C} = \{S(t) \mid S'(t) \leq 0 \quad \forall t\}$, following

---

[1] Appendix: https://hal.archives-ouvertes.fr/hal-01849180/document.

[11]. This leads to a second likelihood term constraining the dynamics of the temporal sources:

$$p(\mathcal{C}|S', \lambda) = (1 + \exp(-\lambda S'(t)))^{-1} \tag{3}$$

We jointly optimize (2) according to priors and constraints, by maximizing the data evidence:

$$\log(p(Y, \mathcal{C}|\sigma, \lambda)) = \log[\int_A \int_S \int_{S'} p(Y|A, S, \sigma)p(\mathcal{C}|S', \lambda)p(A)p(S, S'|\lambda)dA dS dS'] \tag{4}$$

Since this integral is intractable, we tackle the optimization of (4) via stochastic variational inference. Following [10] and [5] we introduce approximations $q_1(A)$ and $q_2(\Omega, W)$ to derive the lower bound:

$$\begin{aligned} \log(p(Y, \mathcal{C}|\sigma, \lambda)) \geqslant \ &\mathbb{E}_{A \sim q_1, (\Omega, W) \sim q_2}[log(p(Y|A, \Omega, W, \sigma))] + \mathbb{E}_{(\Omega, W) \sim q_2}[log(p(\mathcal{C}|\Omega, W, \lambda))] \\ &- \mathcal{D}[q_1(A)||p(A)] - \mathcal{D}[q_2(\Omega, W)||p(\Omega, W)] \end{aligned} \tag{5}$$

where $\mathcal{D}$ refers to the Kullback-Leibler divergence.

We specify the approximated distribution of the spatial activation maps q_1 such that $q_1(A) = \prod_{n=1}^{Ns} \mathcal{N}(\mu_n, \Sigma(\alpha, \beta))$. To introduce spatial correlations in the maps we choose $\Sigma_{i,j}(\alpha, \beta) = \alpha \exp(-||u_i - u_j||^2/2\beta)$ to model a smooth decay across voxels with coordinates (u_i, u_j). We follow [5] and [11] to also define a variational lower bound on the constrained GPs parameterizing the temporal processes. Thanks to the proposed framework, (4) can be efficiently optimized by stochastic variational inference through backpropagation. We chose to alternate the optimization between the spatio-temporal parameters and the time-shift. We set λ to the minimum value that gives monotonic sources, while σ was arbitrarily determined from the data. A detailed derivation of the model and lower-bound can be found in the Appendix.

3   Results

3.1   Benchmark on Synthetic Data

We tested the algorithm on synthetic data to assess its ability to separate spatio-temporal sources from mixed data, and to provide a model selection via the variational lower bound. We generated three monotonically increasing functions $S_i(t)$ such that $S_i(t) = 1/(1 + \exp(-t + \alpha_i))$, and three synthetic Gaussian activation maps A_1, A_2, A_3 with a 30×30 resolution, to mimick grey matter brain areas (Figs. 1a and b). The data was generated as $Y_{p,j} = S(t_p)A + \mathcal{E}_j$ over 40 time points t_p, where t_p is uniformly distributed in [0,1]. We sampled 50 images at instants t_p and applied our method. To simulate a pure cross-sectional setting the time associated to each input image was set to zero. Figures 1c and d show the estimated spatio-temporal processes when fitting the model with three latent sources. In Fig. 2, we see that the individual time-shift parameter estimated for each subject correlates with the original time used to generate the data. This means that the algorithm correctly positions each subject on the temporal trajectories.

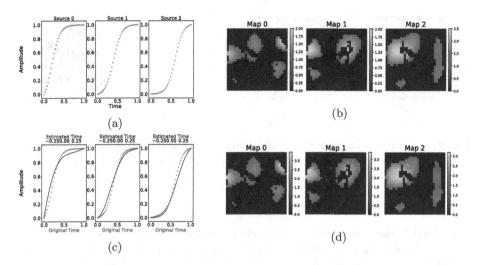

(a)

(b)

(c)

(d)

Fig. 1. (a)–(b) Ground truth temporal and spatial sources. (c) Red: raw temporal sources against the original time axis. Blue: recovered temporal sources against the estimated time scale. (d) Estimated spatial maps. (Color figure online)

To test the model selection, we generated the data as described above using respectively one, two, or three sources over ten folds. For each fold we ran the algorithm looking for one to four sources. Figure 3 shows mean and standard deviation of the lower bound. We observe that when the number of sources is under-estimated the lower bound is higher. When the number of sources is over-estimated, although the lower bound for model selection is more uncertain, by looking at the extracted spatial maps we observe that the additional sources are mainly set to zero or have low weights (see the map of Fig. 3). These experimental results indicate that the optimal number of sources should be selected by inspection of both the lower bound and the extracted spatial sources.

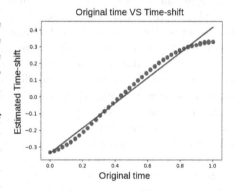

Fig. 2. The red points represent the values of the estimated subjects' time-shift against their associated ground truth value. (Color figure online)

The method was also compared to ICA in a simplified setting by assigning the ground truth parameter t_p beforehand. This simplification is necessary since standard ICA can't be applied when the time associated to each image is unknown. We observed that ICA recovered the spatio-temporal sources, by providing however more noisy estimations than the ones we obtained. This result highlights the importance of the priors and constraints introduced in our method (see Appendix).

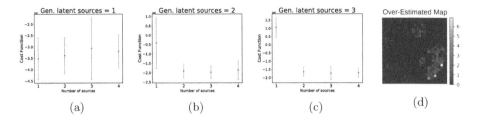

Fig. 3. (a)–(b)–(c): Distribution of the lower bound against the number of fitted sources. (d): 4th extracted spatial map with data generated by 3 latent sources.

3.2 Application on Real Data

Data used in the preparation of this article were obtained from the Alzheimer's Disease Neuroimaging Initiative (ADNI) database (adni.loni.usc.edu). The ADNI was launched in 2003 as a public-private partnership, led by Principal Investigator Michael W. Weiner, MD. For up-to-date information, see www.adni-info.org.

In this section we present an application of the algorithm on real data, using grey matter maps extracted from structural MRI. We selected a cohort of 555 subjects from ADNI composed by 94 healthy controls, 343 MCI, and 118 AD patients. We processed the baseline MRI of each subject to obtain high-dimensional grey matter density maps in a standard space [1]. We extracted the 90×100 middle coronal slice for each patient, to obtain a data matrix Y with dimensions 555×9000, and applied our algorithm looking for three spatio-temporal sources (see Fig. 4). The middle spatial map shows a strong activation of the hippocampus, while the left and right plots show an activation on the temporal lobes, with two similar temporal behaviours, characterized by a less pronounced grey matter loss compared to the hippocampus. More specifically, we observe that the hippocampal trajectory has a strong acceleration in opposition to the other brain areas. This pattern quantified by our model in a pure data-driven manner is compatible with empirical evidence from clinical studies [2]. In Fig. 5 we observe the estimated time of each patient against standard

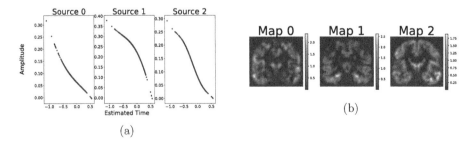

Fig. 4. (a)–(b) Temporal and spatial sources extracted from the data.

volumetric and clinical biomarkers. We see a strong correlation between brain volumetric measures and the estimated time, as well as a non-linear relation in the evolution of ADAS11. The latter result indicates an acceleration of clinical symptoms along the estimated time course.

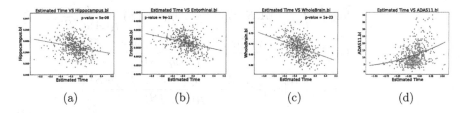

(a)   (b)   (c)   (d)

Fig. 5. Evolution of volumetric and clinical biomarkers along the estimated time.

4   Conclusion

We presented a method for analyzing spatio-temporal data, which provides both independent spatio-temporal processes at stake in AD, and a disease progression scale. Applied on grey matter maps, the model highlights different brain regions affected by the disease, such as the hippocampus and the temporal lobes, along with their differential temporal trajectory. We also show a strong correlation between the estimated disease progression scale and different clinical and volumetric biomarkers. We are currently extending the approach to scale to 3D volumetric images by parallelization on multiple GPUs. The lower bound properties will be also further investigated to better assess its reliability, in order to improve the model comparison. Moreover the method will be extended beyond the cross-sectional application of Sect. 3.2, to account for time-series of brain images, as well as for multimodal imaging biomarkers. Finally we will investigate the use of the approach for prognosis purposes, to provide a data-driven assessment of disease severity in testing patients.

Acknowledgements. This work has been supported by the French government, through the UCA^{JEDI} Investments in the Future project managed by the National Research Agency (ref.n ANR-15-IDEX-01), the grant AAP Santé 06 2017-260 DGA-DSH, and by the Inria Sophia Antipolis - Méditerranée, "NEF" computation cluster.

Appendix

A. Lower bound derivation

In this section we detail the derivation of the lower bound:

$$\log(p(Y, \mathcal{C}|\sigma, \lambda)) = \log[\int_A \int_S \int_{S'} p(Y|A, S, \sigma)p(\mathcal{C}|S', \lambda)p(A)p(S, S'|\lambda)dAdSdS']$$

$$= \log[\int_A \int_S \int_{S'} p(Y|A, S, \sigma)p(\mathcal{C}|S', \lambda)p(A)p(S'|S, \lambda)p(S)dAdSdS']$$

$$(1)$$

If we know S this completely determines S', thus we have $\int p(S'|S,\lambda)dS' = 1$ which gives us:

$$
\begin{aligned}
\log(p(Y,\mathcal{C}|\sigma,\lambda)) &= \log[\int_A \int_S p(Y|A,S,\sigma)p(\mathcal{C}|S',\lambda)p(A)p(S)dAdS] \\
&= \log[\int_A \int_S p(Y|A,S,\sigma)p(\mathcal{C}|S',\lambda)p(A)p(S)\frac{q_1(A)q_2(S)}{q_1(A)q_2(S)}dAdS] \\
&= \log[\mathbb{E}_{A\sim q_1, S\sim q_2}[\frac{p(Y|A,S,\sigma)p(\mathcal{C}|S',\lambda)p(A)p(S)}{q_1(A)q_2(S)}]] \\
&\geqslant \mathbb{E}_{A\sim q_1, S\sim q_2}[\log[\frac{p(Y|A,S,\sigma)p(\mathcal{C}|S',\lambda)p(A)p(S)}{q_1(A)q_2(S)}]]
\end{aligned}
$$
(2)

This is obtained thanks to Jensen's inequality. Finally this leads us to:

$$
\begin{aligned}
\mathbb{E}_{A\sim q_1, S\sim q_2}[\log[\frac{p(Y|A,S,\sigma)p(\mathcal{C}|S',\lambda)p(A)p(S)}{q_1(A)q_2(S)}]] = &\ \mathbb{E}_{A\sim q_1, S\sim q_2}[\log[p(Y|A,S,\sigma)]] \\
&+ \mathbb{E}_{S\sim q_2}[\log(P(\mathcal{C}|S',\lambda))] \\
&- \mathcal{D}[q_1(A|Y)||p(A)] \\
&- \mathcal{D}[q_2(S|Y)||p(S)]
\end{aligned}
$$
(3)

In the Method section we introduced the approximation $q_1(A) = \prod_{n=1}^{Ns} \mathcal{N}(\mu_n, \Sigma(\alpha,\beta))$. The covariance matrix is shared by all the spatial processes which gives us the set of spatial parameters:

$$
\psi = \{\mu_n, n \in [1, Ns], \alpha, \beta\}
$$
(4)

Following [5] we introduce for each GP two vectors, Ω_n with a prior $p(\Omega_n) = \mathcal{N}(0, \frac{1}{l_n}I)$ for each element and W_n with a prior $p(W_n) = \mathcal{N}(0, I)$, such that $S_n(t) = \Phi(t\Omega_n)W_n$. Where Φ is chosen to obtain a RBF kernel as explained in [5]. We define the approximated distributions $q_3(W_n) = \prod_j \mathcal{N}(m_{n,j}, s_{n,j}^2)$ and $q_4(\Omega_n) = \prod_j \mathcal{N}(\alpha_{n,j}, \beta_{n,j}^2)$ of $p(W_n)$ and $p(\Omega_n)$. Using these approximations and following [5], we can derive a lower bound for S with the same technique than above. We have the set of temporal parameters:

$$
\theta = \{m_n, s_n, \alpha_n, \beta_n, l_n, n \in [1, Ns]\}
$$
(5)

Now we can obtain every term of (3). The Kullback-Leibler of a multivariate Gaussian has a closed-from:

$$
\mathcal{D}[q_1(A|X)||p(A)] = \frac{1}{2}\sum_{n=1}^{Ns} Tr(\Sigma) + \mu_n^T\mu_n - F - \log[det(\Sigma)]
$$
(6)

Using the factorized form of q_2 and the fact that the different Gaussian processes are independent from each other we can write:

$$\mathcal{D}[q_2(S|X)||p(S)] = \sum_{n=1}^{Ns} \mathcal{D}[q_3(W_n)|p(W_n)] + \mathcal{D}[q_4(\Omega_n)|p(\Omega_n)] \quad (7)$$

Since the approximations q_3 and q_4 and their respective priors are normally distributed we have an analytic formula for both Kullback-Leibler divergences.

$$\mathcal{D}[q_3(W_n)|p(W_n)] = \frac{1}{2} \sum_j s_{n,j}^2 + \mu_{n,j}^2 - 1 - log(s_{n,j}^2) \quad (8)$$

$$\mathcal{D}[q_4(\Omega_n)|p(\Omega_n)] = \frac{1}{2} \sum_j \beta_{n,j}^2 l_n + \alpha_{n,j}^2 l_n - 1 - log(\beta_{n,j}^2 l_n) \quad (9)$$

As in [10] we employ the reparameterization trick to have an efficient way of sampling the expectations of (3). Thus we have:

- $W_{n,j} = m_{n,j} + s_{n,j} * \epsilon_{n,j}$
- $\Omega_{n,j} = \alpha_{n,j} + \beta_{n,j} * \zeta_{n,j}$
- $A_n = \mu_n + \Sigma_n^{\frac{1}{2}} * \kappa_n$

Which gives us:

$$\mathbb{E}_{A\sim q_1, S\sim q_2}[log(p(Y|A, S, \sigma))] = \mathbb{E}_{\epsilon,\zeta,\kappa}[log(p(Y|m, s, \alpha, \beta, \mu, \Sigma, \sigma))] \quad (10)$$
$$\mathbb{E}_{S\sim q_2}[log(p(C|S', \lambda))] = \mathbb{E}_{\epsilon,\zeta}[log(p(C|m, s, \alpha, \beta, \lambda))] \quad (11)$$

Where $\epsilon_{n,j} \sim \mathcal{N}(0, 1)$, $\zeta_{n,j} \sim \mathcal{N}(0, 1)$ and $\kappa_n \sim \mathcal{N}(0, I)$.

B. Kronecker factorization

Here we detail how to split the covariance matrix in a Kronecker product of three matrices along each spatial dimensions. We have:

$$\Sigma_{i,j}(\alpha, \beta) = \alpha \exp(-\frac{||u_i - u_j||^2}{2\beta}) \quad (12)$$

We can use the separability properties of the exponential to decompose the covariance between two locations $u_i = (x_i, y_i, z_i)$ and $u_j = (x_j, y_j, z_j)$:

$$\Sigma_{i,j}(\alpha, \beta) = \alpha \exp(-\frac{(x_i - x_j)^2}{2\beta}) \exp(-\frac{(y_i - y_j)^2}{2\beta}) \exp(-\frac{(z_i - z_j)^2}{2\beta}) \quad (13)$$

So Σ can be decomposed into the Kronecker product of 1D processes:

$$\Sigma = \Sigma_x \otimes \Sigma_y \otimes \Sigma_z \quad (14)$$

Allowing us to deal with large-size matrices.

C. Comparison with ICA

We performed a comparison of our algorithm with ICA on a similar example than in Sect. 3.1. However the data was generated in a simplifed setting since ICA can't be applied when the time associated to each image is unknown. To do so we assigned the ground truth parameter t_p beforehand. The goal was to compare the separation performances of both our algorithm and ICA, on data generated with three latent spatio-temporal processes. In Fig. 6 we observe that the sources estimated by ICA are more noisy and uncertain than the ones estimated by our method, highlighting the performances of our algorithm in terms of sources separation.

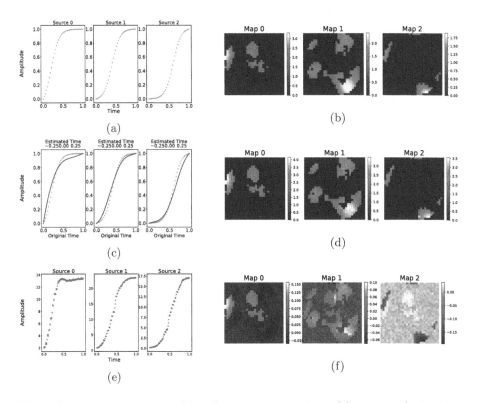

Fig. 6. First row: raw sources. Second row: sources estimated by our method. Third row: sources estimated by ICA.

D. ADNI

Data collection and sharing for this project was funded by the Alzheimer's Disease Neuroimaging Initiative (ADNI) and DOD ADNI. ADNI is funded by the National Institute on Aging, the National Institute of Biomedical Imaging and

Bioengineering, and through generous contributions from the following: Abb-Vie, Alzheimer's Association; Alzheimer's Drug Discovery Foundation; Araclon Biotech; BioClinica, Inc.; Biogen; Bristol-Myers Squibb Company; CereSpir, Inc.; Cogstate;Eisai Inc.; Elan Pharmaceuticals, Inc.; Eli Lilly and Company; EuroImmun; F. Hoffmann-La Roche Ltd. and its affiliated company Genentech, Inc.; Fujirebio; GE Healthcare; IXICO Ltd.; Janssen Alzheimer Immunotherapy Research & Development, LLC.; Johnson & Johnson Pharmaceutical Research & Development LLC.; Lumosity; Lundbeck; Merck & Co., Inc.; Meso Scale Diagnostics, LLC.; NeuroRx Research; Neurotrack Technologies;Novartis Pharmaceuticals Corporation; Pfizer Inc.; Piramal Imaging; Servier; Takeda Pharmaceutical Company; and Transition Therapeutics. The Canadian Institutes of Health Research is providing funds to support ADNI clinical sites in Canada. Private sector contributions are facilitated by the Foundation for the National Institutes of Health (www.fnih.org). The grantee organization is the Northern California Institute for Research and Education, and the study is coordinated by the Alzheimer's Therapeutic Research Institute at the University of Southern California. ADNI data are disseminated by the Laboratory for Neuro Imaging at the University of Southern California.

References

1. Ashburner, J.: A fast diffeomorphic image registration algorithm. NeuroImage **38**(1), 95–113 (2007)
2. Bateman, R.J., et al.: Clinical and biomarker changes in dominantly inherited Alzheimer's disease. New Engl. J. Med. **367**(9), 795–804 (2012). pMID: 22784036
3. Bilgel, M., et al.: Temporal trajectory and progression score estimation from voxelwise longitudinal imaging measures: application to amyloid imaging. Inf.Process. Med. Imaging **24**, 424–436 (2015)
4. Calhoun, V.D., et al.: A review of group ICA for fMRI data and ICA for joint inference of imaging, genetic, and ERP data. Neuroimage **45**(1 Suppl), S163–S172 (2009)
5. Cutajar, K., et al.: Random feature expansions for deep Gaussian processes. In: Precup, D., Teh, Y.W. (eds.) Proceedings of the 34th International Conference on Machine Learning. Proceedings of Machine Learning Research, vol. 70, pp. 884–893. PMLR, International Convention Centre, Sydney, Australia, 06–11 August 2017
6. Donohue, M.C., et al.: Estimating long-term multivariate progression from short-term data. Alzheimer's Dementia **10**(Suppl. 5), S400–S410 (2014)
7. Hyvärinen, A., Oja, E.: Independent component analysis: algorithms and applications. Neural Netw. **13**, 411–430 (2000)
8. Jack, C.R.: Hypothetical model of dynamic biomarkers of the Alzheimer's pathological cascade. Lancet Neurol **9**(1), 119–128 (2010)
9. Jedynak, B.M.: A computational neurodegenerative disease progression score: method and results with the Alzheimer's disease neuroimaging initiative cohort. Neuroimage **63**(3), 1478–1486 (2012)
10. Kingma, D.P., Welling, M.: Auto-encoding variational bayes. CoRR abs/1312.6114 (2013)

11. Lorenzi, M., Filippone, M.: Constraining the dynamics of deep probabilistic models. In: Dy, J., Krause, A. (eds.) Proceedings of the 35th International Conference on Machine Learning, Proceedings of Machine Learning Research, vol. 80, pp. 3233–3242. PMLR, Stockholmsmässan, Stockholm, Sweden, 10–15 July 2018

12. Lorenzi, M., et al.: Probabilistic disease progression modeling to characterize diagnostic uncertainty: application to staging and prediction in Alzheimer's disease. NeuroImage (2017)

13. Marinescu, R.V., et al.: A vertex clustering model for disease progression: application to cortical thickness images. In: Niethammer, M., et al. (eds.) IPMI 2017. LNCS, vol. 10265, pp. 134–145. Springer, Cham (2017). https://doi.org/10.1007/978-3-319-59050-9_11

14. Rahimi, A., Recht, B.: Random features for large-scale kernel machines. In: Platt, J.C. (ed.) Advances in Neural Information Processing Systems, vol. 20, pp. 1177–1184. Curran Associates Inc., New York (2008)

Multi-channel Stochastic Variational Inference for the Joint Analysis of Heterogeneous Biomedical Data in Alzheimer's Disease

Luigi Antelmi[1(✉)], Nicholas Ayache[1], Philippe Robert[2,3], Marco Lorenzi[1], and for the Alzheimer's Disease Neuroimaging Initiative[*]

[1] University of Côte d'Azur, Inria Sophia Antipolis, Epione Research Project, Nice, France
luigi.antelmi@inria.fr
[2] University of Côte d'Azur, CoBTeK, Nice, France
[3] Centre Memoire, CHU of Nice, Nice, France

Abstract. The joint analysis of biomedical data in Alzheimer's Disease (AD) is important for better clinical diagnosis and to understand the relationship between biomarkers. However, jointly accounting for heterogeneous measures poses important challenges related to the modeling of heterogeneity and to the interpretability of the results. These issues are here addressed by proposing a novel multi-channel stochastic generative model. We assume that a latent variable generates the data observed through different channels (*e.g.*, clinical scores, imaging) and we describe an efficient way to estimate jointly the distribution of the latent variable and the data generative process. Experiments on synthetic data show that the multi-channel formulation allows superior data reconstruction as opposed to the single channel one. Moreover, the derived lower bound of the model evidence represents a promising model selection criterion. Experiments on AD data show that the model parameters can be used for unsupervised patient stratification and for the joint interpretation of the heterogeneous observations. Because of its general and flexible formulation, we believe that the proposed method can find various applications as a general data fusion technique.

[*]Data used in preparation of this article were obtained from the Alzheimer's Disease Neuroimaging Initiative (ADNI) database (adni.loni.usc.edu). As such, the investigators within the ADNI contributed to the design and implementation of ADNI and/or provided data but did not participate in analysis or writing of this report. A complete listing of ADNI investigators can be found at: http://adni.loni.usc.edu/wp-content/uploads/how_to_apply/ADNI_Acknowledgement_List.pdf. A detailed list of funding actors can be found in the *Acknowledgments* section of the *Supplementary Material*.

Electronic supplementary material The online version of this chapter (https://doi.org/10.1007/978-3-030-02628-8_2) contains supplementary material, which is available to authorized users.

D. Stoyanov et al. (Eds.): MLCN 2018/DLF 2018/iMIMIC 2018, LNCS 11038, pp. 15–23, 2018.
https://doi.org/10.1007/978-3-030-02628-8_2

1   Introduction

Physicians investigate their patients' status through various sources of information that in this work we call *channels*. For Alzheimer's Disease (AD), for example, the anamnestic questionnaire, genetic tests and various imaging modalities are channels providing specific, complementary, and sometimes overlapping views on the patient's state [3, 7].

Tackling a complex disease like AD requires to establish a link between heterogeneous data channels. However, simple univariate correlation analyses are limited in modeling power, and are prone to false positives when the data dimension is high. To overcome the limitations of mass-univariate analysis, more advanced methods, such as Partial Least Squares (PLS), Reduced Rank Regression (RRR), or Canonical correlation analysis (CCA) [5] have successfully been applied in biomedical research [12], along with multi-channel [8, 13] and non-linear [1, 6] variants.

A common drawback of standard multivariate methods is that they are not generative. Indeed, their formulation consists in projecting the observations in a latent lower dimensional space in which they exhibit certain desired characteristics like maximum correlation (CCA), maximum covariance (PLS), minimum regression error (RRR); however these methods are limited in providing information on how this latent representation is expressed in the observations [4]. Moreover, techniques for model comparison should be applied to select the best number of dimensions for the latent representation and avoid overfitting. While cross-validation is the standard model validation procedure, this requires holding-out data from the original dataset, and thus leading to data loss at the training stage.

We need generative models that can actually describe the direct influence of the latent space on the observations, and model selection techniques leveraging solely on training data. *Bayesian-CCA* [11] actually goes in this direction: it is a generative formulation of the CCA defined on a latent variable that captures the shared variation between data channels. Moreover, the Bayesian formulation allows the use of probabilistic model comparison, without recurring to cross-validation. However, *Bayesian-CCA* may not scale well to large dimensions and several channels.

In this work we aim at addressing the current methodological limitations in multi-channel analysis. By leveraging on the recent developments on performing efficient approximate *Variational Inference* in Bayesian modeling in an efficient way, we propose a novel multi-channel stochastic generative model for the joint analysis of multi-channel heterogeneous data. Our hypothesis is that a latent variable $\mathbf{z}$ generates the heterogeneous data $\mathbf{x}_1, \ldots, \mathbf{x}_C$ observed through different channels C. In this work we propose an efficient way to estimate jointly the latent variable distribution and the data likelihood $p(\mathbf{x}_1, \ldots, \mathbf{x}_C | \mathbf{z})$, and we also investigate a mean for Bayesian model selection. Our work generalizes the *Variational Autoencoder* [10] and the *Bayesian-CCA*, making possible to jointly model multiple channels simultaneously and efficiently.

The next sections of this paper are organized as follows. In Sect. 2 we present the derivation of the multi-channel variational model and we describe a possible implementation with Gaussian distributions parametrized by linear functions. In Sect. 3 we apply our method on a synthetic dataset, as well as on a real multi-channel Alzheimer's disease dataset, to test the descriptive and predictive properties of the model. In the last section we provide our discussions and conclusions. Further experimental tests are provided in the *Supplementary Material*[1].

2   Method

2.1   Multi-channel Variational Inference

Let $\mathbf{x} = \{\mathbf{x}_c\}_{c=1}^C$ be a single observation of a set of C channels, where each $\mathbf{x}_c \in \mathbb{R}^{d_c}$ is a d_c-dimensional vector. Also, let $\mathbf{z} \in \mathbb{R}^l$ denote the l-dimensional latent variable commonly shared by each $\mathbf{x}_c$. We propose the following generative process:

$$\mathbf{z} \sim p(\mathbf{z})$$
$$\mathbf{x}_c \sim p(\mathbf{x}_c|\mathbf{z}, \boldsymbol{\theta}_c) \qquad \text{for} \quad c \quad \text{in} \quad 1 \dots C \tag{1}$$

where $p(()\,z)$ is a prior distribution for the latent variable and $p(\mathbf{x}_c|\mathbf{z}, \boldsymbol{\theta}_c)$ is a likelihood distribution for the observations conditioned on the latent variable. We assume that the likelihood functions belong to a distribution family $\mathcal{P}$ parametrized by $\boldsymbol{\theta}_c$. When the distributions are Gaussians parametrized by linear transformations, the model is equivalent to the *Bayesian-CCA* (*cf.* [11], Eq. 3). In the scenario depicted so far, solving the inference problem allows the discovery of the common latent space from which the observed data in each channel is generated. The solution to the inference problem is given by deriving the posterior $p(\mathbf{z}|\mathbf{x}_1, \dots, \mathbf{x}_C, \boldsymbol{\theta}_1, \dots, \boldsymbol{\theta}_C)$, that is not always computable analytically. In this case, *Variational Inference* [2] can be applied to compute an approximate posterior. In our setting, variational inference is carried out by introducing probability density functions $q(\mathbf{z}|\mathbf{x}_c, \boldsymbol{\phi}_c)$ that are on average as close as possible to the true posterior in terms of Kullback-Leibler divergence:

$$\underset{q \in \mathcal{Q}}{arg\,min} \; \mathbb{E}_c\left[\mathcal{D}_{KL}\big(q(\mathbf{z}|\mathbf{x}_c, \boldsymbol{\phi}_c) \,\|\, p(\mathbf{z}|\mathbf{x}_1, \dots, \mathbf{x}_C, \boldsymbol{\theta}_1, \dots, \boldsymbol{\theta}_C)\big)\right] \tag{2}$$

where the approximate posteriors $q(\mathbf{z}|\mathbf{x}_c, \boldsymbol{\phi}_c)$ belong to a distribution family $\mathcal{Q}$ parametrized by $\boldsymbol{\phi}_c$, and represent the view on the latent space that can be inferred from each channel $\mathbf{x}_c$. Practically, solving the objective in Eq. 2 allows to use on average every $q(\mathbf{z}|\mathbf{x}_c, \boldsymbol{\phi}_c)$ to approximate the true posterior distribution. It can be shown that the maximization of the model evidence $p(\mathbf{x}_1, \dots, \mathbf{x}_C)$ is equivalent to the optimization of the evidence lower bound $\mathcal{L}(\boldsymbol{\theta}, \boldsymbol{\phi}, \mathbf{x})$:

$$\mathcal{L}(\boldsymbol{\theta}, \boldsymbol{\phi}, \mathbf{x}) = \frac{1}{C} \sum_{c=1}^{C} \mathbb{E}_{q(\mathbf{z}|\mathbf{x}_c, \boldsymbol{\phi}_c)} \underbrace{\left[\sum_{i=1}^{C} \ln p(\mathbf{x}_i|\mathbf{z}, \boldsymbol{\theta}_c)\right]}_{\text{cross-reconstruction of all } \mathbf{x}_i \text{ from } \mathbf{x}_c} - \mathcal{D}_{KL}\big(q(\mathbf{z}|\mathbf{x}_c, \boldsymbol{\phi}_c) \,\|\, p(\mathbf{z})\big)$$

$$= \underbrace{\ln p(\mathbf{x}_1, \dots, \mathbf{x}_C)}_{\text{Evidence}} - \underbrace{\mathbb{E}_c\left[\mathcal{D}_{KL}\big(q(\mathbf{z}|\mathbf{x}_c, \boldsymbol{\phi}_c) \,\|\, p(\mathbf{z}|\mathbf{x}_1, \dots, \mathbf{x}_C, \boldsymbol{\theta}_1, \dots, \boldsymbol{\theta}_C)\big)\right]}_{\geq 0} \tag{3}$$

---

[1] https://hal.inria.fr/hal-01844733.

It can be shown that maximizing $\mathcal{L}\left(\boldsymbol{\theta}, \boldsymbol{\phi}, \mathbf{x}\right)$ is equivalent to solving the objective in Eq. 2 (*cf. Sup. Mat.*). Moreover, being the lower bound linked to the data evidence up to a positive constant, Eq. 3 allows to test $\mathcal{L}\left(\boldsymbol{\theta}, \boldsymbol{\phi}, \mathbf{x}\right)$ as a surrogate measure of $p\left(\mathbf{x}_1, \ldots, \mathbf{x}_C\right)$ for Bayesian model selection. This formulation is valid for any distribution family $\mathcal{P}$ and $\mathcal{Q}$, and the complete derivation of Eq. 3 is in the *Sup. Mat.*

Comparison with Variational Autoencoder (VAE). Our model extends the VAE [10]: the novelty is in the cross-reconstruction term labeled in Eq. 3. In case $C = 1$ the model collapses to a VAE. In the case $C > 1$ the cross-term forces each channel to the joint decoding of the other channels. For this reason, our model is different from a stack of VAEs. The dependence between encoding and decoding across channels stems from the joint approximation of the posteriors (Formula (2)).

Optimization of the Lower Bound. The optimization starts with a random initialization of the generative parameters $\boldsymbol{\theta}$ and the variational parameters $\boldsymbol{\phi}$. The expectation in the first row of Eq. 3 can be computed by sampling from the variational distributions $q\left(\mathbf{z}|\mathbf{x}_c, \boldsymbol{\phi}_c\right)$ and, when the prior and the variational distributions are Gaussians, the Kullback-Leibler term can be computed analytically (*cf.* [10], Appendix 2.A). The maximization of $\mathcal{L}\left(\boldsymbol{\theta}, \boldsymbol{\phi}, \mathbf{x}\right)$ with respect to $\boldsymbol{\theta}$ and $\boldsymbol{\phi}$ is efficiently carried out through minibatch stochastic gradient descent implemented with the backpropagation algorithm. For each parameter, adaptive learning rates are computed with *Adam* [9].

2.2   Gaussian Linear Case

Model (1) is completely general and can account for complex non-linear relationships modeled, for example, through deep neural networks. However, for simplicity of interpretation, and validation purposes, in the next experimental section we will restrict our multi-channel variational framework to the *Gaussian Linear Model*. This is a special case, analogous to Bayesian-CCA, where the members of the generative family $\mathcal{P}$ and variational family $\mathcal{Q}$ are Gaussians parametrized by linear transformations. The parameters of these transformations are thus optimized by maximizing lower bound. We define the members of the generative family $\mathcal{P}$ as Gaussians whose first moments are linear transformations of the latent variable $\mathbf{z}$, and the second moments are parametrized by a diagonal covariance matrix, such that $p\left(\mathbf{x}_c|\mathbf{z}, \boldsymbol{\theta}_c\right) = \mathcal{N}\left(\mathbf{x}_c|\mathbf{G}_c^{(\mu)}\mathbf{z}, diag(\mathbf{g}_c^{(\sigma)})\right)$, where $\mathbf{G}_c^{(\mu)} \in \mathbb{R}^{d_c \times l}$ and $\mathbf{g}_c^{(\sigma)} \in \mathbb{R}^{d_c}$. The elements of $\boldsymbol{\theta}_c = \{\mathbf{G}_c^{(\mu)}, \mathbf{g}_c^{(\sigma)}\}$ are the generative parameters to be optimized for every channel. We also define the members of variational family $\mathcal{Q}$ to be Gaussians whose moments are linear transformation of the observations, such that $q\left(\mathbf{z}|\mathbf{x}_c, \boldsymbol{\phi}_c\right) = \mathcal{N}\left(\mathbf{x}_c|\mathbf{V}_c^{(\mu)}\mathbf{x}_c, diag(\mathbf{V}_c^{(\sigma)}\mathbf{x}_c)\right)$ where $\mathbf{V}_c^{(\mu)} \in \mathbb{R}^{l \times d_c}$ and $\mathbf{V}_c^{(\sigma)} \in \mathbb{R}^{l \times d_c}$. The elements of $\boldsymbol{\phi}_c = \{\mathbf{V}_c^{(\mu)}, \mathbf{V}_c^{(\sigma)}\}$ are the variational parameters to be optimized for every channel.

3   Experiments

In this section we illustrate the performance of the method extensively tested on a large scale synthetic dataset, and we provide a real case example by jointly analyzing multimodal brain imaging and clinical scores in AD data. Further experimental tests are provided in *Sup. Mat.*

3.1   Experiments on Linearly Generated Synthetic Datasets

Data Generation Procedure. Datasets $\mathbf{x} = \{\mathbf{x}_c\}$ with $c = 1 \ldots C$ channels where created as $\mathbf{x}_c = \mathbf{G}_c \mathbf{z} + snr^{-1/2}\boldsymbol{\epsilon}$, where $\mathbf{z} \sim \mathcal{N}(\mathbf{0}; \mathbf{I}_l)$ and $\boldsymbol{\epsilon} \sim \mathcal{N}(\mathbf{0}; \mathbf{I}_{d_c})$. snr is the signal-to-noise ratio and $\mathbf{G}_c$ is the linear generative law initialized as $\mathbf{G}_c = diag\left(\mathbf{R}_c \mathbf{R}_c^T\right)^{-1/2} \mathbf{R}_c$ where, for every channel c, $\mathbf{R}_c \in \mathbb{R}^{d_c \times l}$ is a random matrix with l orthonormal columns (*i.e.*, $\mathbf{R}_c^T \mathbf{R}_c = \mathbf{I}_l$). It's easy to demonstrate that the diagonal elements of the covariance matrix of $\mathbf{x}_c$ are inversely proportional to snr, *i.e.*, $diag\left(\mathbb{E}\left[\mathbf{x}_c \mathbf{x}_c^T\right]\right) = (1 + snr^{-1})\mathbf{I}_{d_c}$. Scenarios where generated by varying one-at-a-time the dataset attributes (*e.g.*, noise level, number of observations, ...), as listed in *Sup. Mat.* We fitted several instances of the model specified in Sect. 2.2, changing each time the number of fitted latent dimensions, for a total of 40 000 experiments.

Results. At convergence, the loss function (negative lower bound) has a minimum when the number of fitted latent dimension corresponds to the number of the latent dimensions used to generate the data, as depicted in Fig. 1a. When increasing the number of fitted latent dimensions, a sudden decrease of the loss (elbow effect) is indicative that the true number of latent dimensions has been found. In the *Sup. Mat.* we show also that the elbow effect becomes more pronounced with increasing the number of data channels. Ambiguity in identifying the elbow, instead, may rise for high-dimensional data channels. In these cases, increasing the sample size or the data quality in terms of snr can make the elbow point more noticeable.

Concerning the estimation of the ground truth parameters and data reconstruction, we observed that the performance of the model increases with higher snr, sample size, and number of channels (*Sup. Mat.*; moreover we notice that the error made in ground truth data recovery with multi-channel information is systematically lower than the one obtained with a single-channel decoder (Fig. 1b).

3.2   Application to Clinical and Medical Imaging Data in AD

Data Preparation. Data used in the preparation of this article were obtained from the Alzheimer's Disease Neuroimaging Initiative (ADNI) database (adni.loni.usc.edu). The ADNI was launched in 2003 as a public-private partnership, led by Principal Investigator Michael W. Weiner, MD. For up-to-date information, see www.adni-info.org.

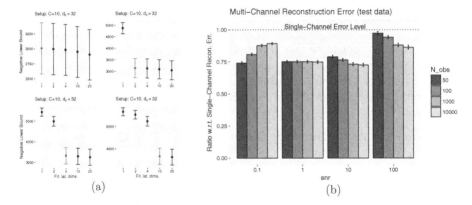

Fig. 1. (a) Negative lower bound (NLB) on the synthetic training set computed at convergence for all the scenarios. Each bar shows mean ± s.e. of $N = 80$ total experiments as a function of the number of fitted latent dimensions. Red bars represents experiments where the number of true and fitted latent dimensions coincide. (b) Ratio between Multi- *vs* Single-Channel reconstruction error computed as mean squared error from the ground truth test data. (Color figure online)

We fit our model with linear parameters to clinical imaging channels acquired on 504 subjects. The clinical channel is composed of six continuous variables generally recorded in memory clinics (age, mini-mental state examination, adas-cog, cdr, faq, scholarity); the three imaging channels are structural MRI (gray matter only), functional FDG-PET, and Amyloid-PET, each of them composed by continuous measures averaged over 90 brain regions mapped in the AAL atlas [14]. Raw data from the imaging channels where coregistered in a common geometric space, and visual quality check was performed to exclude registration errors. Data was centered and standardized across dimensions. Model selection was carried out by comparing the lower bound for several fitted latent dimensions.

Results. As depicted in Fig. 2a, we found that model selection through the lower bound identifies in a range around 16 the number of latent dimensions that optimally describe the observations. When fixing 16 latent dimensions, in one of them (Fig. 2c) subjects appear stratified by disease status, an information that was not directly introduced ahead. For each model, the classification accuracy in predicting the disease status was assessed through split-half cross-validation linear discriminant analysis on the latent variables (Fig. 2b). Maximum accuracy for disease classification occurs at 16 and 32 latent dimensions, an optimum location also identified through the lower bound. Figure 3 shows the generative parameters ϕ_c of the four channels associated to the latent dimension shown in Fig. 2c. The generative parameters describe a plausible relationship between this latent dimension and the heterogeneous observations in the data channels.

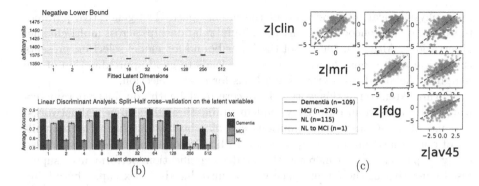

(a)

(b)

(c)

Fig. 2. Modeling results on ADNI data. (a) The negative lower bound has a minimum when fitting 16 latent dimensions. (b) Classification performance of the models: maximum accuracy for classes identification occurs with 16 and 32 lat. dims., in agreement with (a). (c) Pairwise representations of one latent dimension (out of 16) inferred from each of the four data channel. Although the optimization is not supervised to enforce clustering, subjects appear stratified by disease classes.

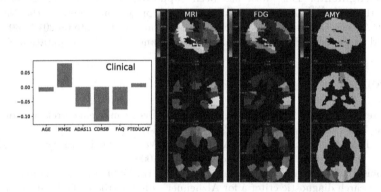

Fig. 3. Generative parameters $\phi_c^{(\mu)}$ of the four channels associated to the latent dimension in Fig. 2c. The clinical parameters are age, mini-mental state examination (mmse), adas-cog (adas11), cdr-sb, faq, scholarity (pteducat); The generative parameters describe a plausible relationship between the latent variable and the heterogeneous observations in the ADNI dataset, coherently with the research literature on Alzheimer's Disease (e.g. low amyloid deposition, high mmse, high scholarity, low cdr, etc.).

4   Discussion and Conclusion

We presented a multi-channel stochastic framework based on a probabilistic generative formulation. The performance of our multi-channel model was shown in the case of Gaussian distributions with moments parametrized by linear functions. In the real case scenario of AD modeling, the model allowed the unsupervised stratification of the latent variable by disease status, providing evidence for a physiological interpretation of the latent space. The generative parameters can therefore encode clinically meaningful relationships across multi-channel observations. Although the use of the lower bound for model selection presents theoretical limitations [2], we found that it leads to good approximation of the marginal likelihood, thus providing a basis for model selection.

Future extension of this work will concern model with non-linear parameterization of the distributions, easily implementable through deep neural networks. The use of non-Gaussian distributions can also be tested. Given the scalability of our variational model, application to high resolution images may be also easily implemented. To increase the model classification performance, supervised clustering of the latent space will be introduced, for example, by adding an appropriate cost function to the lower bound. Also, introducing sparsity to remove redundancies may ease the identification and interpretation of the most informative parameters. Lastly, due to the general formulation, the proposed method can find various applications as a general data fusion technique, not limited to the biomedical research area.

Acknowledgments. This work has been supported by: the French government, through the UCA^{JEDI} Investments in the Future project managed by the National Research Agency (ref. n. ANR-15-IDEX-01); the grant AAP Santé 06 2017-260 DGA-DSH, and by the Inria Sophia Antipolis - Méditerranée, "NEF" computation cluster.

References

1. Andrew, G., Arora, R., Bilmes, J., Livescu, K.: Deep canonical correlation analysis. Proc. Mach. Learn. Res. **28**(3), 1247–1255 (2013)
2. Blei, D.M., Kucukelbir, A., McAuliffe, J.D.: Variational inference: a review for statisticians (2016). http://arxiv.org/abs/1601.00670
3. Dubois, B., Feldman, H.H., Jacova, C., Hampel, H., Molinuevo, J.L., et al.: Advancing research diagnostic criteria for Alzheimer's disease: the IWG-2 criteria. Lancet Neurol. **13**(6), 614–629 (2014)
4. Haufe, S., Meinecke, F., Görgen, K., Dähne, S., Haynes, J.D., et al.: On the interpretation of weight vectors of linear models in multivariate neuroimaging. Neuroimage **87**, 96–110 (2014)
5. Hotelling, H.: Relations between two sets of variates. Biometrika **28**(3/4), 321 (1936)
6. Huang, S.Y., Lee, M.H., Hsiao, C.K.: Nonlinear measures of association with kernel canonical correlation analysis and applications. J. Stat. Plan. Inference **139**(7), 2162–2174 (2009)
7. Jack, C.R., Bennett, D.A., Blennow, K., Carrillo, M.C., Dunn, B., et al.: NIA-AA Research Framework: toward a biological definition of Alzheimer's disease. Alzheimer's Dement. **14**(4), 535–562 (2018)
8. Kettenring, J.R.: Canonical analysis of several sets of variables. Biometrika **58**(3), 433–451 (1971)
9. Kingma, D.P., Ba, J.: Adam: a method for stochastic optimization. CoRR abs/1412.6 (2014). http://arxiv.org/abs/1412.6980

10. Kingma, D.P., Welling, M.: Auto-encoding variational bayes. In: Proceedings 2nd International Conference on Learning Representations, ICLR 2014, December 2014. http://arxiv.org/abs/1312.6114
11. Klami, A., Seppo, V., Kaski, S.: Bayesian canonical correlation analysis. J. Mach. Learn. Res. **14**, 965–1003 (2013)
12. Liu, J., Calhoun, V.D.: A review of multivariate analyses in imaging genetics. Front. Neuroinform. **8**, 29 (2014)
13. Luo, Y., Tao, D., Ramamohanarao, K., Xu, C., Wen, Y.: Tensor canonical correlation analysis for multi-view dimension reduction. IEEE Trans. Knowl. Data Eng. **27**(11), 3111–3124 (2015)
14. Tzourio-Mazoyer, N., Landeau, B., Papathanassiou, D., Crivello, F., Etard, O., et al.: Automated anatomical labeling of activations in SPM using a macroscopic anatomical parcellation of the MNI MRI single-subject brain. Neuroimage **15**(1), 273–289 (2002)

Visualizing Convolutional Networks
for MRI-Based Diagnosis
of Alzheimer's Disease

Johannes Rieke[1,2(✉)], Fabian Eitel[1], Martin Weygandt[1], John-Dylan Haynes[1],
and Kerstin Ritter[1]

[1] Charité – Universitätsmedizin Berlin, corporate member of Freie Universität Berlin,
Humboldt-Universität zu Berlin, and Berlin Institute of Health (BIH); Bernstein
Center for Computational Neuroscience, Berlin Center for Advanced Neuroimaging,
Department of Neurology, and Excellence Cluster NeuroCure, Berlin, Germany
johannes.rieke@gmail.com
[2] Technical University Berlin, Berlin, Germany

Abstract. Visualizing and interpreting convolutional neural networks
(CNNs) is an important task to increase trust in automatic medical
decision making systems. In this study, we train a 3D CNN to detect
Alzheimer's disease based on structural MRI scans of the brain. Then,
we apply four different gradient-based and occlusion-based visualization
methods that explain the network's classification decisions by highlighting relevant areas in the input image. We compare the methods qualita-
tively and quantitatively. We find that all four methods focus on brain
regions known to be involved in Alzheimer's disease, such as inferior and
middle temporal gyrus. While the occlusion-based methods focus more
on specific regions, the gradient-based methods pick up distributed relevance patterns. Additionally, we find that the distribution of relevance
varies across patients, with some having a stronger focus on the temporal
lobe, whereas for others more cortical areas are relevant. In summary,
we show that applying different visualization methods is important to
understand the decisions of a CNN, a step that is crucial to increase
clinical impact and trust in computer-based decision support systems.

Keywords: Alzheimer · Visualization · MRI · Deep learning · CNN
3D · Brain

1 Introduction

Alzheimer's disease (AD) is the main cause of dementia in the elderly. It is
symptomatically characterized by loss of memory and other intellectual abilities to such an extent that it affects daily life. Long before memory problems
occur, microscopic changes related to cell death take place and slowly progress
over time. Radiologically, neurodegeneration is the hallmark of AD, starting
in the temporal lobe and then spreading all over the brain. However, since all

© Springer Nature Switzerland AG 2018
D. Stoyanov et al. (Eds.): MLCN 2018/DLF 2018/iMIMIC 2018, LNCS 11038, pp. 24–31, 2018.
https://doi.org/10.1007/978-3-030-02628-8_3

brains from elderly people are affected by atrophy, it is a difficult task (even for experienced radiologists) to discriminate normal age-related atrophy from AD-mediated atrophy.

In this context, machine learning models provide great potential to capture even slight tissue alterations. State-of-the-art models for image classification are convolutional neural networks (CNNs), which have recently been applied to medical imaging data for various use cases [5], including AD detection. The key idea behind CNNs is inspired by the mechanism of receptive fields in the primate visual cortex: Local convolutional filters and pooling operations are applied successively to extract regional information. In contrast to traditional machine learning-based approaches, CNNs do not rely on hand-crafted features but find meaningful representations of the input data during training.

Although CNNs deliver good classification results, they are difficult to visualize and interpret. In medical decision making, however, it is critical to explain the behavior of a machine learning model and let medical experts verify the diagnosis. A number of visualization methods have been suggested that highlight regions in an input image with strong influence on the classification decision [9–12]. Such heatmaps constitute the basis for understanding and interpreting machine learning models, optimally together with clinicians.

In this work, we compare four visualization methods (sensitivity analysis, guided backpropagation, occlusion and brain area occlusion) on a 3D CNN, which was trained to classify structural MRI scans of the brain into AD patients and normal elderly controls (NCs).

2   Related Work

2.1   Alzheimer Classification

A number of machine learning models have been applied to Alzheimer detection. Some use traditional approaches with hand-crafted features, while most recent papers employ deep convolutional networks. For an overview, we refer the reader to Table 1 in Khvostikov et al. [3]. We identified three studies that use a model and training procedure similar to ours (i.e. 3D CNN, full-brain structural MRI scans, AD/NC classification) [2,4,8]: In contrast to our study, Payan et al. [8] and Hosseini-Asl et al. [2] pretrain their convolutional layers with an unsupervised autoencoder. Korolev et al. [4] train from scratch, but use more complex networks. The CNN architecture in our study is partly inspired by a model in Khvostikov et al. [3], even though they only train on images of the hippocampus.

2.2   Visualization Methods

A range of visualization methods for CNNs have recently been developed. While some methods aim to find prototypes for classes, we only use methods that explain the CNN decision for a specific sample [9,10,12] (see methodological details below). We found two studies that apply visualization methods to AD classification in a similar way as we do: Korolev et al. [4] employ the occlusion method on a deep CNN. While they show similar results like ours (focus on hippocampus and ventricles), they do not compare different visualization methods or analyze the relevance distribution in detail. Yang et al. [11] use a segmentation-based occlusion (similar to our brain area occlusion), but reach inconclusive results. To the best of our knowledge, this is the first study that comprehensively compares different visualization methods on CNNs for AD detection.

3   Methods

3.1   Data

For this study we used structural MRI data of patients with Alzheimer's disease (AD) and normal controls (NC) from phase 1 of the Alzheimer's Disease Neuroimaging Initiative (ADNI, http://adni.loni.usc.edu/) who were included in the "MRI collection - Standardized 1.5T List - Annual 2 year". For each subject, this data collection offers structural MRI scans of the full brain for up to three time points (screening, 12 and 24 months; sometimes multiple scans per visit). We excluded scans with mild cognitive disorder (MCI) and two scans for which our preprocessing pipeline failed. In total, our dataset comprises 969 individual scans (475 AD, 494 NC) from 344 subjects (193 AD, 151 NC).

All scans were acquired with 1.5 T scanners at various sites and had undergone gradient non-linearity, intensity inhomogeneity and phantom-based distortion correction. We downloaded T1-weighted MPRAGE scans and non-linearly registered all images to a 1 mm isotropic ICBM template using ANTs (http://stnava.github.io/ANTs/), resulting in volumes of $193 \times 229 \times 193$.

For training, we split this dataset using 5-fold cross validation. The split is performed on the level of patients to prevent the network from seeing images of the same patient during training and testing. For the visualization methods, we used a fixed split with 30 AD and 30 NC patients in the test set.

3.2   Model

Our model consists of four convolutional layers (with filter size $3 \times 3 \times 3$ and 8/16/32/64 feature maps) and two fully-connected layers (128/64 neurons; this architecture is inspired by a model in Khvostikov et al. [3]). We apply batch normalization and pooling after each convolution and dropout of 0.8 before the first fully-connected layer. The network has two output neurons with softmax activation. We train with cross-entropy loss and the Adam optimizer (learning rate 0.0001, batch size 5) for 20 epochs. Before feeding the brain scans to the network, we remove the skull and normalize each voxel to have mean 0 and standard deviation 1 across the training set.

3.3 Visualization Methods

In this section, we briefly review the four visualization methods we used in this study (see also the review by Montavon et al. [6]). All of these methods produce a heatmap over the input image, which indicates the relevance of image pixels for the classification decision. PyTorch implementations of all visualization methods will be made available at http://github.com/jrieke/cnn-interpretability.

Sensitivity Analysis (Backpropagation) [9]. The gradient of the network's output probability w.r.t. the input image is calculated. For a given image pixel, this gradient describes how much the output probability changes when the pixel value changes. In neural networks, the gradient can be easily computed via the backpropagation algorithm, which is used for training. As relevance score, we take the absolute value of the gradient.

Guided Backpropagation [10]. This method is a modified version of sensitivity analysis, in which the negative gradients are set to 0 at ReLU layers during the backward pass. This is equivalent to a combination of Backpropagation and Deconvolution and leads to more focused heatmaps. As above, we take the absolute value of the gradient as the relevance score.

Occlusion [12]. A part of the image is occluded with a black or gray patch and the network output is recalculated. If the probability for the target class decreases compared to the original image, this image region is considered to be relevant. To get a relevance heatmap, we slide the patch across the image and plot the difference between unoccluded and occluded probability (for AD or NC). We use a patch of size $40 \times 40 \times 40$ with value 0.

Brain Area Occlusion. This method is a modification of occlusion, in which we occlude an entire brain area based on the Automated Anatomical Labeling atlas (AAL, http://www.gin.cnrs.fr/en/tools/aal-aal2/). This method was inspired by a segmentation-based visualization in Yang et al. [11]. As for occlusion, we report the difference between unoccluded and occluded probability (for AD or NC).

4 Results

4.1 Classification

Using 5-fold cross-validation, our network achieves a classification accuracy of 0.77 ± 0.06 and ROC AUC of 0.78 ± 0.04 (both mean $\pm$ standard deviation). This is comparable to recent studies for other convolutional networks [2,4,8]. For example, Korolev et al. [4], who use a similar model and training procedure, achieve a similar accuracy of 0.79 ± 0.08, but with a better ROC AUC of 0.88 ± 0.08. Please note that our focus was on the different visualization methods and not on optimizing the network.

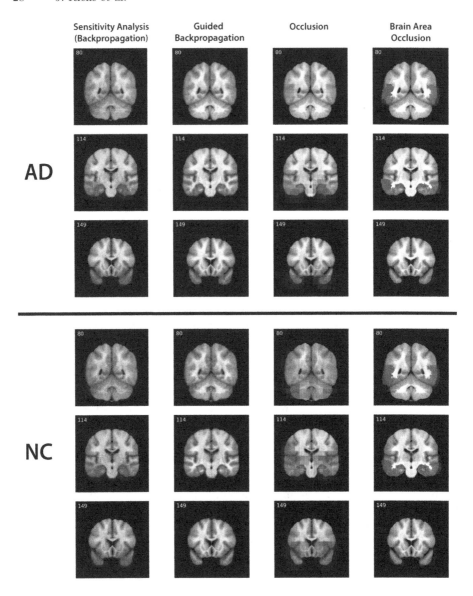

Fig. 1. Relevance heatmaps for all visualization methods, averaged over AD (top) and NC (bottom) samples in the test set. Red indicates relevance, i.e. a red area was important for the network's classification decision. Numbers indicate slice positions (out of 229 coronal slices). (Color figure online)

Table 1. Most relevant brain areas per visualization method, averaged over AD (top) and NC (bottom) samples in the test set. Values in brackets give fraction of summed relevance in this brain area, divided by the summed relevance in the whole brain.

	Sensitivity analysis (Backpropagation)	Guided backpropagation	Occlusion	Brain area occlusion
AD	TemporalMid (6.1%) TemporalInf (5.9%) Fusiform (4.6%) CerebelumCrus1 (3.8%)	TemporalMid (7.0%) TemporalInf (5.7%) FrontalMid (4.2%) Fusiform (3.9%)	TemporalMid (12.1%) TemporalInf (9.2%) Fusiform (6.2%) ParaHippocampal (5.4%)	TemporalMid (29.7%) TemporalInf (14.8%) TemporalSup (4.4%) Hippocampus (4.1%)
NC	TemporalMid (6.1%) TemporalInf (5.8%) Fusiform (4.5%) CerebelumCrus1 (3.8%)	CerebelumCrus1 (4.6%) TemporalMid (4.5%) TemporalInf (4.5%) FrontalMid (4.1%)	TemporalMid (6.2%) TemporalSup (4.9%) CerebelumCrus1 (4.9%) Insula (4.7%)	TemporalMid (20.4%) TemporalInf (12.8%) Fusiform (7.2%) TemporalSup (6.2%)

4.2   Relevant Brain Areas

Figure 1 shows relevance heatmaps for all visualizations methods, averaged over AD and NC samples in the test set. Since there is no ground truth available for such heatmaps, we validate our results by focusing on specific brain areas that were associated with AD in the medical literature. We identified the most relevant brain areas for each visualization method by summing the relevance in each area (according to the AAL atlas). Table 1 lists the four most relevant brain areas for each method, again averaged over AD and NC samples.

For both AD and NC patients, we can see that the main focus of the network is on the temporal lobe, especially its medial part. This brain area, containing the hippocampus and other structures associated with memory, has been empirically linked to AD [1]. The hippocampus itself is usually one of the earliest areas affected by AD [7]. In our experiments, we observe some relevance on the hippocampus, but usually the whole area around it is crucial for the network's decision. This may be explained by the fact that our samples contain only advanced forms of the disease.

In addition to temporal regions, we observe some relevance attributed to other areas across the brain (especially in the gradient-based visualization methods). We find that the distribution of relevance varies between patients: Some brains have strong relevance in the temporal lobe, while in others, the cortex plays a crucial role.

Lastly, we note that the heatmaps for AD and NC samples are quite similar. This makes sense, given that the network should focus on the same regions to detect presence or absence of the disease. Some differences between AD and NC can be found for the occlusion method. We speculate that this might be an artifact of our specific setting (the network might confuse the occlusion patch with brain atrophy, increasing the probability for AD in some brain areas).

4.3   Differences Between Visualization Methods

Although all visualization methods focus on similar brain areas, we can spot some differences: Occlusion and brain area occlusion are more focused on specific regions, while relevance in the gradient-based methods seems more distributed. Obviously, the occlusion-based approaches cannot deal with large areas of distributed relevance (e.g. in the cortex), because these areas will never be covered up completely by the occlusion patch. Therefore, we recommend to apply gradient-based instead of occlusion-based visualization methods for use cases where the relevance is presumably distributed across the input image. Moreover, we find that brain area occlusion is indeed a very natural approach for our context, but it suffers from the fact that only one brain region is covered up at a time. In our case, this leads to very high relevance for the temporal lobe, but hardly any relevance for other brain structures.

To compare the visualization methods quantitatively, we computed Euclidean distances between the average heatmaps ($\sqrt{\sum_i (A_i - B_i)^2}$ for heatmaps A, B and voxels i), as shown in Table 2. In accordance with the visual impression, we find that the gradient-based methods are relatively similar to each other (i.e. low Euclidean distance). The only method that deviates strongly from all other methods is brain area occlusion, which (as stated above) only attributes relevance to a few image regions.

Table 2. Euclidean distance between relevance heatmaps (averaged over all AD/NC samples in the test set) in 10^{-4}.

	Sensitivity analysis (Backpropagation)	Guided backpropagation	Occlusion	Brain area occlusion
Sensitivity analysis (Backpropagation)	0.00/0.00	4.09/4.36	5.15/4.09	11.48/9.04
Guided backpropagation	4.09/4.36	0.00/0.00	6.47/5.83	11.36/9.80
Occlusion	5.15/4.09	6.47/5.83	0.00/0.00	11.16/9.66
Brain area occlusion	11.48/9.04	11.36/9.80	11.16/9.66	0.00/0.00

5   Conclusion

In this study, we trained a 3D CNN for Alzheimer classification and applied various visualization methods. We show that our CNN indeed focuses on brain regions associated with AD, in particular the medial temporal lobe. This is consistent across all four visualization methods. Interestingly, the distribution of relevance varies between patients, with some having a stronger focus on the temporal lobe, whereas for others more cortical areas were involved. We hope that explaining classifier decisions in this way can pave the way for machine learning models in critical areas like medicine and will increase trust in computer-based decision support systems. Our results also show that the visualization methods differ in their explanations. Therefore, we strongly recommend to compare available visualization methods for a specific application area and not "blindly" trust the results of one method.

For future research, we identified three main areas: First, other visualization methods [6] could be implemented and compared to our results. Second, future studies might apply our workflow to preconditions of Alzheimer's disease, i.e. mild cognitive impairment, and measures of clinical disability. Third, it would be interesting to produce some form of ground truth for the relevance heatmaps, e.g. by implementing simulation models that control for the level of separability or location of differences.

References

1. Frisoni, G.B., Fox, N.C., Jack, C.R., Scheltens, P., Thompson, P.M.: The clinical use of structural MRI in Alzheimer disease. Nat. Rev. Neurol. **6**(2), 67–77 (2010). https://www.nature.com/articles/nrneurol.2009.215
2. Hosseini-Asl, E., Gimel'farb, G., El-Baz, A.: Alzheimer's disease diagnostics by a deeply supervised adaptable 3D convolutional network. Front. Biosci. **23**, 584–596 (2018). https://doi.org/10.1109/ICIP.2016.7532332. http://arxiv.org/abs/1607.00556
3. Khvostikov, A., Aderghal, K., Benois-Pineau, J., Krylov, A., Catheline, G.: 3D CNN-based classification using sMRI and MD-DTI images for Alzheimer disease studies. arXiv preprint (2018). http://arxiv.org/abs/1801.05968
4. Korolev, S., Safiullin, A., Belyaev, M., Dodonova, Y.: Residual and plain convolutional neural networks for 3D brain MRI classification. In: ISBI (2017). https://doi.org/10.1109/ISBI.2017.7950647
5. Litjens, G., et al.: A survey on deep learning in medical image analysis. Med. Image Anal. **42**, 60–88 (2017). https://doi.org/10.1016/j.media.2017.07.005
6. Montavon, G., Samek, W., Müller, K.R.: Methods for interpreting and understanding deep neural networks. Digit. Signal Process. **73**, 1–15 (2018). https://doi.org/10.1016/j.dsp.2017.10.011
7. Mu, Y., Gage, F.H.: Adult hippocampal neurogenesis and its role in Alzheimer's disease. Mol. Neurodegener. **6**(1), 85 (2011). https://doi.org/10.1186/1750-1326-6-85
8. Payan, A., Montana, G.: Predicting Alzheimer's disease: a neuroimaging study with 3D convolutional neural networks. arXiv preprint (2015). https://doi.org/10.1613/jair.301, http://arxiv.org/abs/1502.02506
9. Simonyan, K., Vedaldi, A., Zisserman, A.: Deep inside convolutional networks: visualising image classification models and saliency maps. arXiv preprint (2014). http://arxiv.org/abs/1312.6034
10. Springenberg, J.T., Dosovitskiy, A., Brox, T., Riedmiller, M.: Striving for simplicity: the all convolutional net. In: ICLR (2015). https://doi.org/10.1163/_q3_SIM_00374, http://arxiv.org/abs/1412.6806
11. Yang, C., Rangarajan, A., Ranka, S.: Visual explanations from deep 3d convolutional neural networks for Alzheimer's disease classification. arXiv preprint (2018)
12. Zeiler, M.D., Fergus, R.: Visualizing and understanding convolutional networks. In: Fleet, D., Pajdla, T., Schiele, B., Tuytelaars, T. (eds.) ECCV 2014. LNCS, vol. 8689, pp. 818–833. Springer, Cham (2014). https://doi.org/10.1007/978-3-319-10590-1_53

Finding Effective Ways to (Machine) Learn fMRI-Based Classifiers from Multi-site Data

Roberto Vega[✉] and Russ Greiner

University of Alberta, Edmonton, AB T6G 2R3, Canada
{rvega,rgreiner}@ualberta.ca

Abstract. Machine learning techniques often require many training instances to find useful patterns, especially when the signal is subtle in high-dimensional data. This is especially true when seeking classifiers of psychiatric disorders, from fMRI (functional magnetic resonance imaging) data. Given the relatively small number of instances available at any single site, many projects try to use data from multiple sites. However, forming a dataset by simply concatenating the data from the various sites, often fails, due to batch effects – that is, the accuracy of a classifier learned from such a multi-site datasets, is often worse than of a classifier learned from a single site. We show why several simple, commonly used, techniques – such as including the site as a covariate, z-score normalization, or whitening – are useful only in very restrictive cases. Additionally, we propose an evaluation methodology to measure the impact of batch effects in classification studies and propose a technique for solving batch effects under the assumption that they are caused by a linear transformation. We empirically show that this approach consistently improve the performance of classifiers in multi-site scenarios, and presents more stability than the other approaches analyzed.

Keywords: Multi-site fMRI · Batch effects · Machine learning

1 Introduction

Over the last years, many researchers have been seeking tools that can help with the diagnosis and prognosis of mental health problems. Research groups have used machine learning approaches in the analysis of fMRI data in order to build predictors that can diagnose, for example, attention deficit and hyperactivity disorders, mild cognitive impairment and Alzheimer's disease, schizophrenia, or autism [2]. The reported accuracy of the different tasks varies from chance level to >85%, depending on the task, dataset, features, and learning algorithm used for creating the classifier.

Supported by the Mexican National Council of Science and Technology (CONACYT), Canada's Natural Science and Engineering Research Council (NSERC) and the Alberta Machine Intelligence Institute (AMII).

D. Stoyanov et al. (Eds.): MLCN 2018/DLF 2018/iMIMIC 2018, LNCS 11038, pp. 32–39, 2018.
https://doi.org/10.1007/978-3-030-02628-8_4

One of the main obstacles that limits the usability and generalization capabilities (to new instances) of machine learning approaches is the usually small number of instances (n) of the datasets used to train the models [2]. This is especially problematic when there are a large number of features (p), which might range from a few hundreds to millions depending on the approach, known as "*small n, large p*" [8]. This situation is undesirable because machine learning approaches assume that the training sample is a good approximation of the real distribution of the data, which might not be the case with only a few instances in a high dimensional space.

1.1   Multi-site Data and Batch Effects

In order to mitigate this problem, many researchers use a larger datasets, formed by aggregating fMRI scans obtained at different locations into a single dataset. Unfortunately, inter-scanner variability, possibly caused by field strength of the magnet, manufacturer and parameters of the MRI scanner or radio-frequency noise environments [7], creates a second problem known as *batch effects* [12], which is technical noise that might confound the real biological signal. The main consequence of batch effects in prediction studies is that researchers have observed a *decrease* in classification accuracy on multi-site studies compared with that obtained using a single site [3,12,16].

An underlying assumption of machine learning is that the training set and test set are sampled from the same probability distribution. Because of batch effects, data coming from different sites follow different probability distributions, which might cause the predictors to have a decrease in performance. These discrepancies between the training and test sets are known as dataset shift [14]. This paper focuses on a specific subcase: Let $P_A(X,Y)$ be the joint distribution of the covariates X (the features extracted from the fMRI data) and the label Y (*e.g.*, healthy control or schizophrenia) of scanning site A, and $P_B(X,Y)$ be the corresponding probability distribution for a scanning site B, then $P_A(Y\,|\,X) \neq P_B(Y\,|\,X)$, and $P_A(X) \neq P_B(X)$, but there is a function $g(X)$ such that $P_A(Y\,|\,X) = P_B(Y\,|\,g(X))$, and $P_A(X) = P_B(g(X))$. This concept is exemplified in Fig. 1.

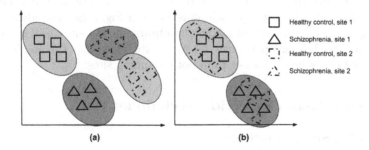

Fig. 1. (a) The dataset sampled from the scanning site 1 follows a different probability distribution than data scanned on site 2. (b) After applying $g(X)$ to the data of site 1 both sites follow the same probability distribution.

The problem of removing batch effects is closely related to that of domain adaptation in the computer vision community [5]. Although some of these approaches have been tested on fMRI data, the performance of classifiers learned from multi-site datasets is in many cases lower than using a single site [16]. The objective of this paper is to analyze some techniques for removing batch effects and the situations where they can be effectively used.

2 Machine Learning and Functional Connectivity Graphs

The standard approach for applying machine learning to fMRI data begins by parcellating the (properly preprocessed) brain volumes into m regions of interest. It then forms a symmetric $m \times m$ pairwise connectivity matrix, whose (i, j) entry each correspond to some measure of statistical dependence between regions i and j, whose upper-triangle is vectorized into a vector of length $p = \frac{1}{2}m(m-1)$.

The vectors corresponding to each of the n subjects in the training set are arranged into a matrix X of dimensions $n \times p$. Similarly, a vector Y of length n, contains the labels of X. Finally, this labeled training data (X, Y) is given to a learning algorithm, that produces the final classifier. A detailed description of this procedure can be found elsewhere [15,16].

A critical aspect in assessing the impact of batch effects in classification studies, as well as the effectiveness of the techniques applied to removed them, is the methodology used to measure the performance of the classifiers. Some studies pool together the data from the different sites and then randomly split the data into a training and test set, while others use the data from $(r-1)$ sites for training and the rth site for testing [1,6,11,12]. The first approach might mask the influence of batch effects because it artificially makes the distribution of the training set and test set more similar. This is an unrealistic scenario. In a real application, a clinician cares about the performance of the classifier on the patients that s/he is evaluating. The second approach is more realistic, but also more complicated. If there is indeed a function $g(X)$ that makes $P_1(Y \mid X) = P_2(Y \mid g(X))$ then we need information from both scanning sites to learn it.

We propose a third evaluation scenario: Fix the test set to be a specific subset of the data from site A. Then consider two training sets: just the remaining instances from site-A versus those remaining site-A instances *and also the instances from site B*. This approach, illustrated in Fig. 2, has the advantage of identifying if there is a benefit of mixing data from different sites, or if it is better to train one classifier independently for every site. Note that this methodology requires having a labeled dataset from both scanning sites.

3 Batch Effects Correction Techniques

3.1 Adding Site as Covariate

This technique involves augmenting each instance with its site information – encoded as a 1-hot-encoding. (That is, using r additional bit features, where the

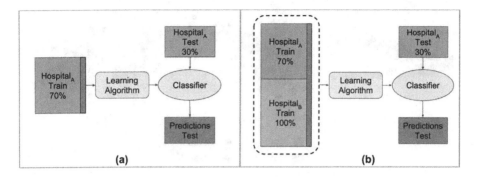

Fig. 2. Evaluating a classifier in single site (a) and multi-site (b) scenarios.

j^{th} feature is 1 if that instance comes from the jth site, and the other features here are 0.)

When using a linear classifier, this method assumes that the only difference between sites is in the threshold that we use to classify an instance as belonging to one class, or another. If we assume that the decision function for one site is given by $w^T x = 0$, where the x vector represent the features and w is the vector of the coefficients (or weights) of the features, then the decision function for a second site is given by $w^T x + c = 0$. This method is effective when the batch effect is caused by a translation (adding a constant) to each instance of the dataset, but it will be ineffective otherwise. Figure 3(a) shows an illustration of this case. Note how the learned decision boundary is appropriate for one of the sites (red), but suboptimal for the other (blue). Note that this technique forces both decision boundaries to have the same slope, and only the bias changes.

3.2 Z-Score Normalization

This approach modifies the probability distribution of the features extracted from *both* sites, A and B, by making the values of each individual feature, for each site, zero-mean with unit variance – i.e., for each site, for the ith feature, subject its mean (for that site), and divide by its empirical standard deviation (for that site). Using this technique, only the marginals are the same in both sites, but the covariance structure is not. Applying this "Z-score normalization" to the data from every scanning site independently, will effectively remove batch effects caused by translation and scaling of features (see Appendix A.1). However it fails with more complex transformations, such as rotations or linear transformations in general; see Fig. 3(b). Note that this scaling and translation is in the feature space, and so it is different to the affine transformations that are corrected during the preprocessing stage (which are applied in the coordinate space).

3.3 Whitening

Whitening is a linear transformation that can be viewed as a generalization of the z-score normalization. Besides making the mean of every feature equal

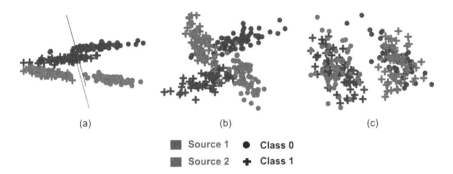

(a)   (b)   (c)

Source 1   ● Class 0
Source 2   ✚ Class 1

Fig. 3. Examples of linear transformation where the methods fail. (a) Including site as covariate, (b) z-score normalization, (c) whitening. (Color figure online)

to zero and its variance equal to one, it also removes the correlation between features by making the overall covariance matrix the identity matrix. One of the most common procedures to perform this process is *PCA Whitening* [10]. This transformation first rotates the data, in each site, by projecting it into its principal components, and then it scales the rotated data by the square root of its eigenvalues (which represent the variance of each new variable in the PCA space). Applying this whitening transformation to every dataset independently will remove the batch effects caused by a rotation and translation of the datasets, since in this cases the principal components of the different sites will be aligned; see Appendix A.2 for the mathematical derivation. However, since there is no guarantee that the principal components will be aligned in general, it might not work with other linear transformations; see Fig. 3(c).

3.4 Solving Linear Transformations

Note that z-score normalization and whitening solve specific cases of

$$X_B = \alpha X_A + \beta \quad \alpha \in \mathbb{R}^{p \times p}, \quad \beta \in \mathbb{R}^p \tag{1}$$

(corresponding to Eq. 5 in Appendix A.2.) Z-score solves batch effects when the associated matrix α is diagonal, while whitening solves them when α is orthogonal with determinant 1. Nevertheless, both methods fail to solve batch effects for a general matrix α. Note also that the previous approaches did not explicitly compute α and β, but instead, applied a transformation that removed their effects under the specified circumstances. Of course, if we could compute α and β, or even a good approximation $\hat{\alpha}$ and $\hat{\beta}$, we could then solve for any batch effect corresponding to an arbitrary linear transformation.

For any two random vectors X_A and X_B, such that $X_B = \alpha X_A + \beta$:

$$\begin{aligned} \mu_B &= E[X_B] = E[\alpha X_A + \beta] = \alpha E[X_A] + \beta = \alpha \mu_A + \beta \\ \Sigma_B &= COV[X_B] = COV[\alpha X_A + \beta] = \alpha COV[X_A]\alpha^T = \alpha \Sigma_A \alpha^T \end{aligned} \tag{2}$$

Although we can obtain empirical estimates of μ_A, μ_B, Σ_A, Σ_B from the dataset, the problem is in general ill-defined – $i.e.$, there is an infinite number of solutions. Now note that every site includes (at least) two different subpopulations – $e.g.$, healthy controls versus cases (perhaps people with schizophrenia). Each subpopulation has its own mean vector and covariance matrix (μ_A^{HC}, μ_A^{SCZ}, μ_B^{HC}, μ_B^{SCZ}, and $\Sigma_A^{HC}, \Sigma_A^{SCZ}, \Sigma_B^{HC}, \Sigma_B^{SCZ}$). A reasonable assumption is that the batch effects affect both populations in the same way, but by computing the mean and covariance matrix of every population and site independently we are effectively increasing the number of equations available. We can then get an estimate for α and β as follows:

$$\hat{\alpha}, \hat{\beta} \quad = \quad \underset{\alpha,\beta}{\arg\min} \sum_{j \in \{HC,SCZ\}} \sqrt{p}||\mu_B^j - (\alpha\mu_A^j + \beta)||_2 + ||\Sigma_B^j - (\alpha\Sigma_A^j\alpha^T)||_F \quad (3)$$

where p is the dimensionality of the feature set, and $||\cdot||_F$ is the Frobenius norm of a matrix. Note that it is possible to combine data from more than two datasets by finding a linear transformation for every pair of sites.

4   Experiments and Results

4.1   Dataset

We applied the four aforementioned methods to the task of classifying healthy controls and people with schizophrenia using the data corresponding to the Auditory Oddball task to the FBIRN phase II dataset, which is a multisite study developed by the Function Biomedical Informatics Research Network (FBIRN). Keator et $al.$ provides a complete description of the study [9].

After preprocessing the data, we eliminated the subjects who presented head movement greater than the size of one voxel at any point in time in any of the axis, a rotation displacement greater than 0.06 radians, or that did not pass a visual quality control assessment. The original released data contains scans extracted from 6 different scanning sites; however, we only used 4 of them. One of the sites was discarded because it lacked T1-weighted images, which were required as part of our preprocessing pipeline. The second discarded site contained only 6 subjects (5 with schizophrenia) after the quality control assessment, so it was not suitable for our analysis. In summary, we have 21 participants from Site 1, 22 from Site 2, 23 from Site 3 and 23 from Site 4. In all cases, the proportion of healthy controls vs people with schizophrenia is ∼50%.

4.2   Experiments and Results

To obtain the feature vector of every fMRI scan, we used the subset corresponding to the Fronto-Parietal Network for a total of $k = 25$ out of the 264 regions of interest defined by Power et $al.$ [13]. The time series corresponding to every region was simply the average time series of all the voxels inside the region. In

Table 1. Average accuracy after correcting batch effects. The number in entry (i, j) is the accuracy, over instances from the target site i, of the classifier learned by adding all of site j to the training subset of site i. The colored cells indicate results whose difference improves (green) or decrease (red) relative to the single site classification.

	S 1	S 2	S 3	S 4
S 1	62.8	72.3	65.7	67.3
S 2	67.8	66.4	70.0	59.5
S 3	55.0	60.9	58.3	56.9
S 4	62.3	57.8	76.4	71.4

(a) No correction

	S 1	S 2	S 3	S 4
S 1	62.8	70.7	64.7	68
S 2	67.1	66.4	68.1	57.6
S 3	55.7	57.6	58.3	56.9
S 4	67.1	57.6	75.7	71.4

(b) Site as covariate

	S 1	S 2	S 3	S 4
S 1	62.8	64.7	57.6	65.7
S 2	68.5	66.4	67.6	62.3
S 3	48.0	54.0	58.3	58.0
S 4	63.5	56.0	74.0	71.4

(c) Z-score normalization

	S 1	S 2	S 3	S 4
S 1	62.8	55.7	52.8	49.5
S 2	51.6	66.4	52.1	50.4
S 3	54.2	54.2	58.3	53.8
S 4	50.7	47.3	52.6	71.4

(d) Whitening

	S 1	S 2	S 3	S 4
S 1	62.8	65.9	66.4	66.2
S 2	66.6	66.4	67.8	67.8
S 3	49.5	50.2	58.3	51.4
S 4	73.5	72.8	73.5	71.4

(e) Linear transformation

order to obtain the functional connectivity matrix, we computed the Pearson's correlation between the time series of all $\binom{k}{2}$ pairs of regions.

We produced classifiers using a support vector machine (SVM) with linear kernel using the SVMLIB library [4]. The parameters of the SVM were set using cross validation. We applied the batch effect correction techniques previous to merging the datasets into a single training set, and repeated the experiment 15 times with different train/test splits. All the parameters required for the batch effects correction techniques were obtained using only the training sets. Table 1 reports the average accuracy over the 15 rounds.

5   Discussion

In each of the sub-tables in Table 1, the (i, j) entry represent the average accuracy when the training set has instances from the ith and jth site, and the test set has instances only from the ith site. Ideally, all the off-diagonal values should be higher than the diagonal ones; however, this is not the case. In most of the cases we have mixed and inconsistent results. The only method that consistently improves the performance of the classifiers is the one that solves for arbitrary linear transformations (Table 1e). Note that site S3 is an exception, where we do not see any improvement; however, this particular site has a low performance even in the single site scenario. It is likely that the signal in this particular site is too low and cannot be properly detected by the used methods.

These results reinforce the idea that batch effects play predominant role in classification studies, and motivate the need to develop techniques that address them in order to be able to effectively combine multi-site datasets. We can additionally conclude that whitening, z-score normalization and adding the site as covariate are insufficient to solve batch effects in fMRI data. Our method for solving linear transformations is the one who consistently improves the results in a multi-site scenario, indicating that it is a step in the right direction.

References

1. Abraham, A., et al.: Deriving reproducible biomarkers from multi-site resting-state data: an Autism-based example. NeuroImage **147**, 736–745 (2017)
2. Arbabshirani, M.R., Plis, S., Sui, J., Calhoun, V.D.: Single subject prediction of brain disorders in neuroimaging: promises and pitfalls. Neuroimage **145**, 137–165 (2016)
3. Brown, M.R.G., et al.: ADHD-200 Global Competition: diagnosing ADHD using personal characteristic data can outperform resting state fMRI measurements. Front. Syst. Neurosci. **6**, 69 (2012)
4. Chang, C.C., Lin, C.J.: LIBSVM: a library for support vector machines. ACM Trans. Intell. Syst. Technol. **2**, 27:1–27:27 (2011)
5. Csurka, G.: Domain adaptation for visual applications: a comprehensive survey. arXiv preprint arXiv:1702.05374 (2017)
6. Gheiratmand, M., et al.: Learning stable and predictive network-based patterns of schizophrenia and its clinical symptoms. NPJ Schizophr. **3**, 22 (2017)
7. Greve, D.N., Brown, G.G., Mueller, B.A., Glover, G., Liu, T.T.: A survey of the sources of noise in fMRI. Psychometrika **78**(3), 396–416 (2013)
8. Hastie, T., Tibshirani, R., Friedman, J.: The Elements of Statistical Learning. SSS. Springer, New York (2009). https://doi.org/10.1007/978-0-387-84858-7
9. Keator, D.B., et al.: The function biomedical informatics research network data repository. NeuroImage **124, Part B**, 1074–1079 (2016). Sharing the wealth: Brain Imaging Repositories in 2015
10. Kessy, A., Lewin, A., Strimmer, K.: Optimal whitening and decorrelation (2015)
11. Nielsen, J.A., et al.: Multisite functional connectivity MRI classification of autism: ABIDE results. Front. Hum. Neurosci. **7**, 599 (2013)
12. Olivetti, E., Greiner, S., Avesani, P.: ADHD diagnosis from multiple data sources with batch effects. Front. Syst. Neurosci. **6**, 70 (2012)
13. Power, J.D., et al.: Functional network organization of the human brain. Neuron **72**(4), 665–678 (2011)
14. Quinonero-Candela, J., Sugiyama, M., Schwaighofer, A., Lawrence, N.D.: When training and test sets are different: characterizing learning transfer (2012)
15. Richiardi, J., Achard, S., Bunke, H., Van De Ville, D.: Machine learning with brain graphs: predictive modeling approaches for functional imaging in systems neuroscience. IEEE Signal Process. Mag. **30**(3), 58–70 (2013)
16. Vega Romero, R.I.: The challenge of applying machine learning techniques to diagnose schizophrenia using multi-site fMRI data (2017)

First International Workshop on Deep Learning Fails Workshop, DLF 2018

Towards Robust CT-Ultrasound Registration Using Deep Learning Methods

Yuanyuan Sun[1(✉)], Adriaan Moelker[1], Wiro J. Niessen[1,2],
and Theo van Walsum[1]

[1] Erasmus MC, Rotterdam, The Netherlands
y.sun@erasmusmc.nl
[2] Delft University of Technology, Delft, The Netherlands

Abstract. Multi-modal registration, especially CT/MR to ultrasound
(US), is still a challenge, as conventional similarity metrics such as
mutual information do not match the imaging characteristics of ultrasound. The main motivation for this work is to investigate whether a
deep learning network can be used to directly estimate the displacement
between a pair of multi-modal image patches, without explicitly performing similarity metric and optimizer, the two main components in a registration framework. The proposed DVNet is a fully convolutional neural
network and is trained using a large set of artificially generated displacement vectors (DVs). The DVNet was evaluated on mono- and simulated
multi-modal data, as well as real CT and US liver slices (selected from
3D volumes). The results show that the DVNet is quite robust on the
single- and multi-modal (simulated) data, but does not work yet on the
real CT and US images.

Keywords: CT · Ultrasound · Liver · Registration · CNN

1 Introduction

Ultrasound (US) is the preferred imaging modality for image guidance in minimally invasive interventions such as liver tumor ablations, as it is real-time,
portable, safe and cheap. However, tumors are not always clearly visible in US
images. In contrast, they are generally visible in preoperative diagnostic CT
images. Fusion of the diagnostic CT and US volumes thus has great potential to
improve US-guided interventions.

Registration of CT and US is the prerequisite of such image fusion, where
the choice of similarity metric is critical. Mutual Information (MI) is a generic
similarity metric used in multi-modal registration where statistical dependency
between the modalities is exploited [1]. However, the nature of US images, such
as the speckled nature and the dependence of the image on the transducer orientation, does not match well with such a standard metric. A local self-similarity
based metric for multi-modal registration was proposed in [2], which uses the

© Springer Nature Switzerland AG 2018
D. Stoyanov et al. (Eds.): MLCN 2018/DLF 2018/iMIMIC 2018, LNCS 11038, pp. 43–51, 2018.
https://doi.org/10.1007/978-3-030-02628-8_5

similarity of small patches in one modality to estimate a local representation of image structure. This metric may work for CT and US registration, but is computationally quite expensive and thus not really applicable for our use case. Wein *et al.* proposed LC^2, a Correlation Ratio [3] based metric, incorporating both intensity and gradient magnitude information from CT images to simulate US in local patches using linear regression [4]. It can handle CT and US better than other metrics, but there is still room for improvement in its accuracy, capture range and speed.

Since recently, deep learning approaches are used in medical image registration. Wu *et al.* [5] propose to learn features from fixed and moving images with a convolutional stacked auto-encoder (CAE) and then use these learned features to replace hand-crafted features in conventional deformable image registration algorithms. Simonovsky *et al.* [6] use a convolutional neural network to learn a general multi-modal similarity metric which is then used in conventional iterative optimization procedures. Most recently, an unsupervised learning-based deformable registration technique was proposed in [7], which uses a conventional similarity metric as the loss function to train the network. Sokooti *et al.* [8] use multi-scale 3D convolutional neural networks for nonrigid registration between 3D CT data. We were inspired by this work as it does not explicitly perform a similarity metric, which is still an open question for real-time CT-US registration.

In this work, we investigated a deep learning network (DVNet) that directly estimates the displacement between a pair of image patches, without explicitly implementing similarity metric and optimizer, the two main components in the conventional methods. Such an approach could be used as a component in patch-based registration approaches, for example [9].

2   Methods

The proposed DVNet is a fully convolutional neural network, which takes a pair of image patches as input. The output of the DVNet is a displacement vector, specifying the displacement between the centres of the pair of patches.

The DVNet is based on the work of [8], but different from its multi-scale, 4-channel architecture, the DVNet starts with two branches with three convolution layers in each one, in order to extract features in a modality-specific manner. Then, the two branches are concatenated to one branch, followed by 9 more convolution layers. These layers are meant to determine the displacement vector, by combining the features from both input patches. Batch normalization and ReLu activation are used in all layers except the final layer. See Fig. 1 for specific parameters.

The DVNet is trained by optimizing the Mean Absolute Errors (MAE) between the predicted displacement vectors (DVs) and the ground truth, using mini-batch stochastic gradient descent Adam [10]. The MAE is the mean of errors of all the n elements (both x and y) of one prediction: $MAE = \frac{1}{n} \sum_{i=1}^{n} |\hat{p}_i - p|$, where $\hat{p}_i$ defines an element of the output patch and p is the target displacement in x or y.

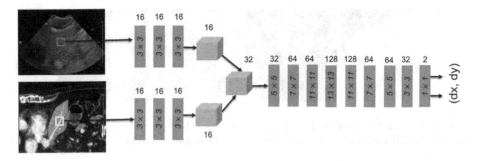

Fig. 1. Network architecture of the DVNet.

3   Data

3.1   Clinical Data

The data we used in the experiments were 30 pairs of CT and US volumes acquired from 18 subjects. Informed consent was obtained from all subjects before image acquisition. The US volume size is $512 \times 378 \times 222$ with voxel size of $0.420 \times 0.387 \times 0.629\,\mathrm{mm}^3$ (acquired using Philips iU22 US system with an X6-1 probe, with the transducer in intercostal and sub-xiphoidal position). The CT slice spacing was $2\,\mathrm{mm}$ (except for one case where slicing spacing was $3\,\mathrm{mm}$), with the pixel size ranging from $0.39\,\mathrm{mm}^2$ to $0.75\,\mathrm{mm}^2$. Each pair of CT and US volumes was annotated by experts with 4 or 5 corresponding landmarks, which served as our reference standard for the rigid alignment of the CT and US.

3.2   Training Data

In this work, the DVs were synthetic, and the training data (fixed patch and moving patch) were generated with these DVs. First, CT and US volumes were rigidly aligned in a same space, using landmarks annotated by clinical experts. Then three or four slice pairs were selected from each pair of the CT and US volumes. To increase the amount and the diversity of the data, 9 more pairs were augmented from each pair of CT and US slices. Specifically, we applied affine transformation (rotation in the range $[-5, 5]$ degrees, shear in $[-15, 15]$ degrees, scale in $[0.9, 1.1]$) and Gaussian noise (with a standard deviation of 5) on the pixel intensities. Subsequently, points were sampled on these slices inside of the liver region, and fixed patches were generated by centering around these points. Then, randomly generated DVs (between -15 and 15 pixels in both x-axis and y-axis) were added to these points to generate the centers of moving patches.

For the single-modality (CT or US) experiments, moving images are the fixed images with Gaussian noise ($\sigma = 5$, scale $= 2$). For the simulated multimodal experiments, fixed patches were from the CT images and moving patches (patch_mov) were created by a linear combination of CT patches and their magnitude gradients, with Gaussian noise ($\sigma = 5$, scale $= 2$), as follows:

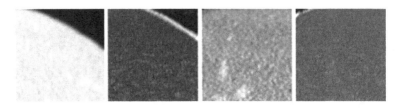

Fig. 2. Examples of simulated multi-modality data. The first and third column are CT patches (fixed) and the other two columns are the simulated data (moving).

$$patch\_mov = \alpha CT + (1 - \alpha)|\nabla CT|, \tag{1}$$

where α was randomly selected within 0 and 1. The data was simulated before training. Figure 2 is an example of the simulated images. The simulation was based on the work of [4], which simulated US from CT, as mentioned in section 1. For CT-US experiments, fixed patches were generated within the fan's region of US and moving patches were within the CT frames.

All the data was divided into a training set of 10 patients, a validation set of 2 patients and a test set of 6 patients. There were about 13,000 patch pairs for training and 2,300 patch pairs for validation. The patch size was 73×73 pixels for all the experiments.

4   Experiments

MeVisLab, Python and PyTorch were used for software development. Experiments were performed on a NVIDIA GeForce GTX 1080 GPU.

Data augmentation was used during the training of DVNet (different for each epoch), with two or three stages. First, the data was randomly flipped both horizontally and vertically. Second, the data was deformed using affine transformation with shearing between -15 and $15°$ and rotation between -5 and $5°$. Third, only for the CT and US experiments, Gaussian noise ($\sigma = 5$, scale $= 2$) was applied and to the CT patches on the pixel intensities. For all experiments, the data was normalized ((x-mean)/std) before training.

In the first experiment, we assessed whether the DVNet can be used to determine displacements. We also used this experiment to determine appropriate settings for some of the hyper-parameters, such as the learning rate and batch size. Subsequently, we investigated the effect of two factors that may hamper convergence of the DVNet on real CT-US image pairs. First, we investigated whether registration of (simulated) multi-modal images is working, and second we investigated the effect of inaccurate ground truth (as the clinical images do not have a perfect ground truth). In the final experiment, we applied the DVNet on the real CT and US, and further investigated factors that may cause the inability of the network to compute displacements for the real imaging data.

4.1 Mono-Modal

The first experiment was performed to assess whether the DVNet approach could be used to estimate the displacements between a pair of image patches. We experimented with different learning rate and batch size, and then set the leaning rate at 0.001 and batch size at 64 for all the subsequent experiments. After 8k iterations, the MAE was around 1 pixel on the test dataset, from which we conclude that the DVNet concept can be used to perform mono-modal registration.

4.2 Multi-modal (Simulated)

Next, in order to investigate if the method is feasible to address multi-modal registration, we simulated multi-modal images as described in Sect. 3.2, with the intension to simulate US from CT. The MAE of the test data was around 1 pixel after 10k iterations. This indicated that the DVNet can deal with the simulated multi-modal data and has the potential to work on the real data.

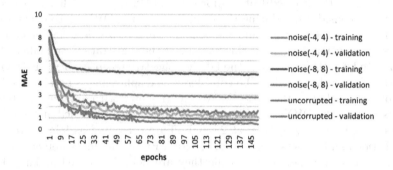

Fig. 3. Loss plots of CT-CT with inaccurate and accurate ground truth.

4.3 Inaccurate Ground Truth

Then, we investigated the influence of inaccurate ground truth, as we do not have an exact ground truth for the real data. We added a random vector (both and in the range [−d, d]) to the target displacement vectors (without changing the images), and used these inaccurate reference displacement vectors for training, while keeping accurate the ground truth in the test set. This experiment was run both for the mono-modal case (CT-CT) and for the (simulated) multi-modal case. Figure 3 contains the loss plots. From the loss plots, it is clear that inaccurate ground truth affects the training loss (increased with the average value of d), but that the testing loss (on uncorrupted data) is only slightly larger than when training with accurate ground truth. From this, we conclude that the DVNet is robust to inaccuracies in the ground truth.

4.4 CT-US

The previous experiments showed that DVNet can be used to get displacement vectors, that it works on the simulated multi-modal images, and that inaccurate ground truth only slightly affects the final results, therefore it still has the potential to work on the real data. However, when we started to train on the real CT-US image pairs, the network failed to converge on the test data (see Fig. 4), which we interpreted as over-fitting.

To address this issue, we experimented with dropout, weight decay and L1/L2 regularization, but it was found that these only delayed the over-fitting and did not actually solve the problem. Initializing the DVNet with pre-trained parameters from the mono- and simulated multi-modal experiment was also investigated, but this did not solve the problem either. We furthermore experimented with first training the DVNet on CT-CT and US-US simultaneously, in order to make the network able to learn features from both modalities. Then, when training on CT-US, the two branches of the DVNet were frozen with parameters from the previous step, as a constant feature extractor. This did not work either, may be because the learned features for CT and US were modality-dependent.

Next, we considered reducing the complexity of the DVNet. To decrease the number of parameters, we could use smaller kernel size, less kernel numbers and less convolution layers, but this may affect the registration capability of the network. To keep the capacity of the DVNet, effective receptive field (ERF) is an important factor that needs to be considered as displacements outside the ERF may not be detected by the network. We experimented with different small ERF to investigate if the DVNet could estimate small DVs while failed at large DVs. Results for a DVNet with ERF at 25 and 39 are presented as joint-histograms of predictions and targets. As we can see from Fig. 5, the measurements gather near the diagonal at small DVs while they are being dispersed at large DVs. This confirms that the size of the ERF is closely related to the capture range of the DVNet. Finally, the DVNet was simplified to the one shown in Fig. 6, at the trade-off between the complexity and the capability. ReLu activation was also replaced by RReLu as it could reduce the risk of over-fitting due to its randomized nature [11]. The new architecture has much less parameters, reduced from several millions to 101,026, with the ERF at 71, almost the size of the input patch. With this new architecture, the MAE of the test data was around 1 mm after 27k and 55k iterations, respectively, for the mono- and the multi-modal images (simulated). The simplified DVNet is also robust to errors on the training reference standard. However, this simplified network can still not converge on the test clinical CT and US data.

Finally, to prevent the over-fitting, we used all the slices from the CT and US volumes and with four more augmentation for each pair of slices to increase the amount of the data. There were about 78,421 patch pairs for training and 14,919 patch pairs for validation. The MAE of the test data in the previous few epochs reduced to a smaller value than before, but then did not go down anymore.

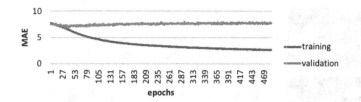

Fig. 4. The loss plot of DVNet on CT and US.

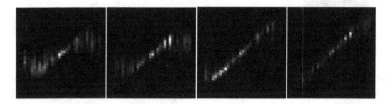

Fig. 5. Joint histograms of predictions and the targets. The ERF is 25 and 39 for the left two and the right two figures, respectively. The first and the third figures are of x-axis displacement and the others are of y-axis. For each figure, both x- and y-axis are in the range of $[-18, 18]$.

5   Discussion and Conclusion

We have presented a patch-based CNN approach for displacement estimation, intended to be used for clinical CT-US registration for image guidance in minimally invasive interventions, called DVNet. We investigated several aspects of the training and application of such a network. Using both the mono-modal and simulated multi-modal images, we demonstrated that the DVNet is capable of accurately computing displacements. We also demonstrated that this learning process is not very sensitive to errors in the training reference standard.

Still, when learning such a network on clinical data, we were not able to get a stable result: after a few iterations over-fitting occurred. Several strategies, such as dropout, weight decay and regularization were employed to address this problem, where we assured that the EFR was sufficiently large such that the displacement vectors can be captured. However, these did not solve the problem.

Our current hypothesis of the failure of training the network is that the relationship between the clinical CT and clinical US images is too difficult to learn with this network. Although it can work on the simulated multi-model data, some effects of US images, such as shadowing caused by high-intensity structures and decreased intensity for deeper structures, are not represented in the simulation. Another reason may be the limited amount of training data. However, we implemented an extensive data augmentation, which generally is very effective in training of CNNs.

In the future, we intend to investigate (deep learning) approaches to simulate CT from US images and vice versa. Such simulated data may be closer to our current multi-modal data, and would serve two purposes. First, it will allow us

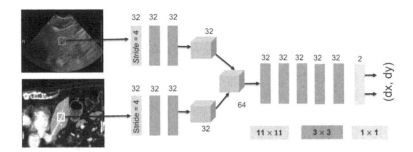

Fig. 6. The simplified DVNet.

to further investigate the multi-modal DVNet approach. Second, such a network may convert the multi-modal registration approach into a mono-modal problem, for which we have shown that DVNet is an effective solution.

In conclusion, the DVNet is a promising method to estimate the local displacement on mono- and multi-modal (simulated) data, and robust to the disturbances in the training reference standard. However, it does not work yet on the clinical CT and US data.

References

1. Pluim, J.P., et al.: Mutual-information-based registration of medical images: a survey. IEEE Trans. Med. Imaging **22**(8), 986–1004 (2003)
2. Heinrich, M.P., et al.: MIND: modality independent neighbourhood descriptor for multi-modal deformable registration. Med. Image Anal. **16**(7), 1423–1435 (2012)
3. Roche, A., Malandain, G., Pennec, X., Ayache, N.: The correlation ratio as a new similarity measure for multimodal image registration. In: Wells, W.M., Colchester, A., Delp, S. (eds.) MICCAI 1998. LNCS, vol. 1496, pp. 1115–1124. Springer, Heidelberg (1998). https://doi.org/10.1007/BFb0056301
4. Wein, W., et al.: Automatic CT-ultrasound registration for diagnostic imaging and image-guided intervention. Med. Image Anal. **12**(5), 577–585 (2008)
5. Wu, G., et al.: Scalable high performance image registration framework by unsupervised deep feature representations learning. IEEE Trans. Biomed. Eng. **63**(7), 1505–1516 (2016)
6. Simonovsky, M., Gutiérrez-Becker, B., Mateus, D., Navab, N., Komodakis, N.: A deep metric for multimodal registration. In: Ourselin, S., Joskowicz, L., Sabuncu, M.R., Unal, G., Wells, W. (eds.) MICCAI 2016. LNCS, vol. 9902, pp. 10–18. Springer, Cham (2016). https://doi.org/10.1007/978-3-319-46726-9_2
7. Vos, B. D., et al.: End-to-end unsupervised deformable image registration with a convolutional neural network. arXiv preprint arXiv:1704.06065 (2017)
8. Sokooti, H., de Vos, B., Berendsen, F., Lelieveldt, B.P.F., Išgum, I., Staring, M.: Nonrigid image registration using multi-scale 3D convolutional neural networks. In: Descoteaux, M., Maier-Hein, L., Franz, A., Jannin, P., Collins, D.L., Duchesne, S. (eds.) MICCAI 2017. LNCS, vol. 10433, pp. 232–239. Springer, Cham (2017). https://doi.org/10.1007/978-3-319-66182-7_27

9. Banerjee, J., et al.: 4D ultrasound tracking of liver and its verification for TIPS guidance. IEEE Trans. Med. Imaging **35**(1), 52–62 (2016)
10. Yang, X., Kwitt, R., Niethammer, M.: Fast predictive image registration. In: Carneiro, G., et al. (eds.) LABELS/DLMIA -2016. LNCS, vol. 10008, pp. 48–57. Springer, Cham (2016). https://doi.org/10.1007/978-3-319-46976-8_6
11. Xu, B., Wang, N., Chen, T., Li, M.: Empirical evaluation of rectified activations in convolutional network. arXiv preprint arXiv:1505.00853 (2015)

To Learn or Not to Learn Features
for Deformable Registration?

Aabhas Majumdar[1]([✉]), Raghav Mehta[1,2], and Jayanthi Sivaswamy[1]

[1] Center for Visual Information Technology (CVIT), IIIT-Hyderabad,
Hyderabad, India
aabhas.majumdar@research.iiit.ac.in
[2] McGill University, Montreal, Canada

Abstract. Feature-based registration has been popular with a variety of features ranging from voxel intensity to Self-Similarity Context (SSC). In this paper, we examine the question of how features learnt using various Deep Learning (DL) frameworks can be used for deformable registration and whether this feature learning is necessary or not. We investigate the use of features learned by different DL methods in the current state-of-the-art discrete registration framework and analyze its performance on 2 publicly available datasets. We draw insights about the type of DL framework useful for feature learning. We consider the impact, if any, of the complexity of different DL models and brain parcellation methods on the performance of discrete registration. Our results indicate that the registration performance with DL features and SSC are comparable and stable across datasets whereas this does not hold for low level features. This shows that when handcrafted features are designed based on good insights into the problem at hand, they perform better or are comparable to features learnt using deep learning framework.

Keywords: Deep learning · Deformable image registration
Brain MRI

1 Introduction

Deformable image registration is critical to tasks such as surgical planning, image fusion, disease monitoring etc. [1]. We focus on application to neuro images where tasks such as Multi-atlas segmentation [2] and atlas construction [3], require registration to handle variations in the shape and size of brain across subjects. Registration entails minimizing a cost function through iterative optimization. Since the cost function quantifies the similarity between the two images to be registered, it plays a crucial role in determining the accuracy of results. Traditional image intensity-based approaches define cost functions based on mutual information, sum of squared difference etc., [4] and use continuous optimization to find the required deformation field. Continuous optimization based methods

---

Aabhas Majumdar and Raghav Mehta are authors with equal contribution.

© Springer Nature Switzerland AG 2018
D. Stoyanov et al. (Eds.): MLCN 2018/DLF 2018/iMIMIC 2018, LNCS 11038, pp. 52–60, 2018.
https://doi.org/10.1007/978-3-030-02628-8_6

require cost functions to be differentiable. With a discrete optimization (DO) formulation, registration has been shown [5, 8] to be more efficient with a 40 to 50-fold reduction in computational time, no loss in accuracy and no requirement of differentiability for cost function. This allows them to use simple cost function like Sum of Absolute Difference (SAD).

Feature based registration was proposed [11] as an improvement over the intensity-based approach. Features that have been explored range from normalized intensity values, edges, geometric moments [11], 3D Gabor attributes [12] and a Modality Independent Neighborhood Descriptor [9]. The natural question to ask is if it is better to learn the features, instead of using hand-crafted ones, since the experience of learning features (using deep networks) for another important problem, namely, segmentation, has been positive [17]. Deep features learnt using unsupervised method [14], has been shown to perform better than traditional features like intensity and edges. A Co-Registration and Co-Segmentation framework [13] has also been proposed using learnt priors on 8 sub-cortical structures and those learnt using a Convolutional Neural Network (CNN) is reported to outperform those learnt with Random-Forrest based classifiers. While these few reports indicate the potential benefit of deep feature learning for registration, it is of interest to gain deeper insights into this specific approach, given the well-established high cost, particularly training overhead, of deep learning which may deter clinical applications with this approach.

In this paper, we seek to gain insights by delving deeper into the issue of feature learning for registration. We attempt to answer the following six questions regarding deep feature learning through extensive experiments on two publicly available datasets. (i) Does complexity of learning architecture matter? (ii) What kind of learning strategy is useful? Supervised or Unsupervised? (iii) What features are better in supervised feature learning? (iv) Does registration accuracy vary with the number of labeled structures in the training data? (v) Does difference in parcellation during training matter? and finally the main question: (vi) To learn or not to learn features for deformable image registration?

To obtain answers to the above questions, a testbed was created for registration using the discrete optimisation framework described in [7] and different Deep Neural Networks (DNNs) were considered for learning the features. The suitability of these features for registration was assessed in terms of the Jaccard Coefficient. At the end, comparison was done against standard low level and high level features like intensity and Self-Similarity Context [6] to assess the requirement of feature learning using deep learning methods.

2   Method

This is a brief overview of the two frameworks which we adopted to setup our testbed and experiments: Discrete Optimization (DO) based registration and Deep Learning based Feature Learning.

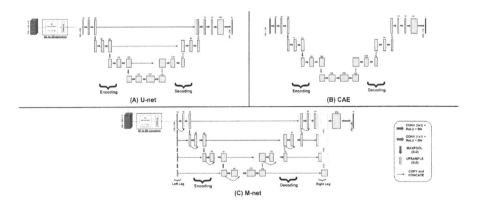

Fig. 1. Three DNNs used for feature learning

2.1   Discrete Optimization

Given a fixed image I_f and moving image I_m the deformation field u required to align I_m to I_f is found via optimizing a cost function $E(u)$. The DO method presented in [7] determines u by minimizing $E(u)$:

$$E(u) = \sum_{\Omega} S(I_f, I_m, u) + \alpha|\nabla u|^2 \tag{1}$$

where Ω is the image patch. The first term S denotes the similarity function between the fixed and warped images while the second term is the regularization term. In our experiments, SAD was chosen as S for the ease of computation.

In DO, the deformation field is only allowed to take values from a quantized set of 3-D displacement for each voxel x. $d \in \{0, \pm q, \pm 2q, ..., \pm l_{max}\}$ Here, q is the quantization step and l_{max} is the maximum range of displacement.

A six-dimensional displacement space volume is created whose each entry is the point-wise similarity cost of translating a voxel x with a displacement d:

$$DSV(x, d) = S(I_f(x), I_m(x + d)) \tag{2}$$

Here, $I_f(x)$ can be simply the voxel intensity or a feature representing the voxel. The displacement field is obtained by winner-takes-all method by selecting the field with the lowest cost for each voxel: $u = \arg\min_d(DSV(d))$. For further, detail into this framework, we refer readers to [7].

2.2   Deep Learning Framework

DNNs are inspired by the biological networks akin to the multilayer perceptron. Basic blocks of DNNs are convolutional layer (2D/3D), maxpooling layer (2D/3D), fully connected layer, dropout layer, activation functions like Rectified Linear Unit (ReLU), tanh, softmax, and batch normalization layer. For a detailed description of these blocks readers are referred to [18].

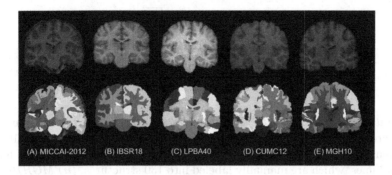

Fig. 2. Coronal slices of brain images (top row) and their manual segmentation (bottom row) of DL training (MICCAI-2012, IBSR18 and LPBA40) and registration testing (CUMC12 and MGH10) datasets.

A Deep Learning framework can be broadly of two types: (i) Supervised and (ii) Unsupervised. The main difference between these two types is that in the former, for any input X, the network tries to predict output Y, which is a class label, while in the latter, the network tries to predict X using the same X as input and in this way network learns something intrinsic about the data without the help of labels generally created by humans.

A CNN [18] is a widely used DNN for supervised learning. CNNs have been used for various tasks such as segmentation and classification. Similarly, Convolutional AutoEncoder (CAE) [19] is a popular framework for unsupervised learning. They are used for varied type of tasks such as learning hidden (or lower dimensional) representation of data, denoising etc.

In our experiments, two CNN architectures, namely U-net [20] and M-net [22] were used for supervised learning. The M-net is an improvement on U-net with added residual and supervision connections. A stack of slices (51) as 3D input is passed through 3D-to-2D converted and then processed by both the architectures to produce segmentation for center slice, as shown in Fig. 1A and C. For more details about these architectures we refer readers to [20] and [22].

CAE is also a DNN which learns useful lower dimensional representation of input from which original input can be generated back with minimal loss of information. The CAE architecture used in our experiment is shown in Fig. 1(B).

3   Experiments and Results

3.1   Datasets Description

In order to ensure thorough evaluation, different datasets were employed for training (the DNNs) versus testing (the registration module). Sample slices of all these datasets are shown in Fig. 2.

Deep Learning Training Datasets: Datasets used for training were chosen according to the diversity in total number of labeled structures and the structure parcellation methods. Details of the chosen datasets are given in Table 1.

Table 1. Deep learning training dataset description

Dataset	# Training volumes	# Validation volumes	# labels	Parcellation type
MICCAI-2012	15	20	135	Whole brain (cortical and non-cortical)
IBSR18	15	3	32	Whole brain (cortical and non-cortical)
LPBA40	30	10	57	Partial brain (mainly cortical)

Registration Testing Datasets: The datasets for testing the registration accuracy were chosen based on their popularity for evaluating registration [10] and variation in structure parcellation methods. *(i) CUMC12:* This dataset has 12 MRI volumes, which are manually labeled into 130 structures. *(ii) MGH10:* This dataset has 10 volumes, segmented into 106 structures. It should be noted that unlike CUMC12 dataset, only cortical structures are marked for this dataset.

3.2   Evaluation Metric

Registration performance was evaluated using mean Jaccard Coefficient (JC). This is the standard evaluation metric employed for comparison of 14 Registration methods in [10]. JC between two binary segmentation A and B is defined as: $JC(A, B) = \frac{|A \cap B|}{|A \cup B|} * 100$. Throughout this paper, we compare registration performance with mean JC which is computed as follows: JC is averaged first across N individual structures for a single volume; then it is averaged across M pairwise registration output. Thus, to evaluate registration performance on the CUMC12 dataset, average JC is found over $N = 130$ structures in a volume and then the average JC is computed for $M = 144$ pairwise registration outputs.

3.3   Implementation Detail

All the DNNs were trained on a NVIDIA K40 GPU, with $12\,GB$ of RAM for 30 epochs. Approximate training time was 3 days. The CNN was trained using Adam Optimizer with following hyper parameters: LR=0.001, β_1=0.9, β_2=0.99, and $\epsilon = 10 * e^{-8}$. LR was reduced by a factor of 10 after 20 epochs. Code was written in Keras Library using Python. The C++ code for DO-based registration made publicly available by the authors of [7] was used. The python code for Deep Learning was integrated in C++ for a seamless implementation.

The effect of intensity variation among training and testing datasets was handled by matching the intensity of all the volumes of testing datasets to that of training dataset volume using Intensity Standardization (IS) [21].

3.4   Feature Learning Experiments and Results

A set of experiments were performed to gain insights into the following six issues in the context of feature-based deformable registration. Registration performance of all these experiments in terms of mean JC is given in Fig. 3.

Role of Complexity of Learning Architecture: Both the U-net and M-net were trained on the MICCAI-2012 dataset and the Segmentation Priors (SP)

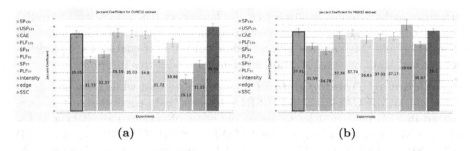

(a)   **(b)**

Fig. 3. Registration performance comparison, in terms of JC, for various deep features on (a) CUMC12 dataset and (b) MGH10 dataset. Here, SP denotes segmentation priors and PLF denotes penultimate layer features.

features were extracted from U-net (USP_{135}) and M-net (SP_{135}) for registration. The mean JC obtained for USP_{135} and SP_{135} on the MGH10 dataset were 35.59 and 37.90 respectively, while they were 31.73 and 35.05 for the CUMC12 dataset. These results indicate that the complexity of architecture does play an important role in feature extraction as registration performance is better for features extracted from a more complex network (M-net).

Supervised or Unsupervised Learning of Features? *:* The M-net and CAE were trained on the MICCAI-2012 dataset. The SP features were extracted from the M-net (SP_{135}) and the Penultimate Layer Features (PLF) from CAE. The mean JC obtained for SP_{135} and CAE were 37.90 and 34.77, respectively for the MGH10 dataset while they were 35.05 and 32.37 for the CUMC12 dataset. Thus, the supervised feature learning appears to be more effective.

Choice of Learnt Features: Registration was done with SP and PLF (hidden layer representation) features separately, after training M-net on the MICCAI-2012 dataset. The obtained mean JC, as shown in Fig. 3, indicates that SP and PLF give comparable performance across both datasets (CUMC12: $SP_{135} =$ 35.05 and $PLF_{135} = 35.19$; MGH10: $SP_{135} = 37.91$ and $PLF_{135} = 37.34$).

Role of the Number of Labeled Structures in Training Data: Available training datasets vary in terms of the number of labeled structures. We can expect the feature learnt on dataset with more structure to differentiate between its neighbouring structures in a better way. In order to understand how this can impact registration, the M-net was trained on two different datasets, namely, MICCAI-2012 (labels: 135) and IBSR18 (labels: 32). The SP and PLF features were extracted from the CNN (M-net) and used in registration. The obtained mean JC for both SP and PLF on both CUMC12 and MGH10 were comparable. (CUMC12: $SP_{135} = 35.05$, $SP_{32} = 35.03$, $PLF_{135} = 35.19$ and $PLF_{32} = 34.9$; MGH10: $SP_{135} = 37.9$, $SP_{32} = 37.73$, $PLF_{135} = 37.34$ and $PLF_{32} = 36.63$) Thus, the features learnt with different number of labeled structures appear to be equally effective for registration. A possible reason for this can be that MICCAI-2012 and IBSR18 datasets have equal number of labels for sub-cortical structures and white matter. However, the former set has a finer level parcellation for cortical structures which essentially encodes spatial position and local information and this may not give added advantage over coarser level parcellation for registration, as registration inherently encodes this information.

Parcellation of Training Dataset: Figure 2 shows that while the whole brain is marked in MICCAI-2012 and IBSR18 datasets, only cortical structures are marked in LPBA40. In order to assess the effect of various parcellation methods, the M-net was trained on the LPBA40 and MICCAI-2012 datasets. SP (SP_{57}, SP_{135}) and PLF (PLF_{57}, PLF_{135}) features were extracted from both. The registration accuracy for both CUMC12 and MGH10 datasets are shown in Fig. 3. It can be seen that there is a drop in JC of approximately 3.33 (9.5%) and 1.33 (3.8%) for SP and PLF respectively, on CUMC12 dataset, relative to the value obtained with features from M-net trained on the MICCAI-2012 dataset ($SP_{135} = 35.05$, $SP_{57} = 31.72$ and $PLF_{135} = 35.19$, $PLF_{57} = 33.86$); whereas on MGH10 dataset, there is only marginal drop in JC of 0.89 (2.3%) and 0.18 (0.4%) for SP and PLF, respectively ($SP_{135} = 37.90$, $SP_{57} = 37.01$ and $PLF_{135} = 37.34$, $PLF_{57} = 37.16$). This can be attributed to the fact that both LPBA40 and MGH10 have only cortical structures marked, while CUMC12 has both cortical and non-cortical structures. Overall, the above results suggest that parcellation method of training dataset should be an important consideration in feature-based registration. Further, it is advisable to train a CNN on a dataset with parcellation for both cortical and sub-cortical structures.

To Learn or Not to Learn Features for Deformable Image Registration? Finally, we turn to the main question of interest: the necessity of feature learning. The registration accuracy of features learnt using M-net was compared against low level features such as intensity, edges as well as a higher level feature, namely, Self-Similarity Context (SSC). The JC values obtained are shown in Fig. 3. Raw intensity feature with SAD as similarity metric has the best performance on MGH10 dataset (39.05) but not on the CUMC12 (29.13) dataset. This is most likely to be due to the persistent voxel intensity variation between the datasets (MGH10 has 32 and CUMC12 has 512 distinct values) despite IS. Interestingly, while both learnt (SP) and high level (SSC) features yield more robust performance across datasets, the latter performs marginally better (CUMC12: SSC $= 35.93$ and $SP_{135} = 35.05$; MGH12: SSC $= 38.1$ and $SP_{135} = 37.9$). Taking the mean JC difference between CUMC12 and MGH10 as a quantifier of robustness, the obtained results ($2.84(SP_{135})$, 2.17 (SSC), 4.72 (edge) and 9.93 (intensity)), indicate that learning *may not* give results superior to hand-crafting of features. SSC is a feature explicitly derived for registration whereas learnt features such as SP are optimised for good segmentation as they are trained on a segmentation dataset.

4   Conclusions

In this paper, the issue of employing learnt (with DNN) features for deformable registration was explored in considerable detail with a set of experiments. Some of the experimental findings such as superiority of supervised features over unsupervised features in terms of robustness is intuitive while others such as accuracy being insensitive to change in the total number of labeled structures during supervised training are counter-intuitive. Our methodology for learning features from a segmentation network was motivated by the widespread practice of assessing registration accuracy indirectly via segmentation as the latter has well defined evaluation metrics. This approach is attractive when *both* problems need to be solved. Learning features (which leads to robust, yet marginally lower performance than SSC) requires considerable computational resources, as one pairwise registration takes 2 mins of CPU + 8 mins of GPU time for feature learnt with DNNs, while it only takes 2–3 mins on CPU for SSC. Recent papers [15,16] have tried to directly learn deformation field for registration instead of features and [15] appears to have slightly better performance than SSC. Taking our findings and based on recent reports, SSC may be a better option in low-resource settings and limited annotated data scenario, especially, if only registration is of interest.

References

1. De Nigris, D., et al.: Fast rigid registration of pre-operative magnetic resonance images to intra-operative ultrasound for neurosurgery based on high confidence gradient orientations. IJCARS **8**(4), 649–661 (2013)
2. Iglesias, J.E., Sabuncu, M.R.: Multi-atlas segmentation of biomedical images: a survey. MedIA **24**(1), 205–219 (2015)
3. Joshi, S., et al.: Unbiased diffeomorphic atlas construction for computational anatomy. NeuroImage **23**, S151–S160 (2004)
4. Penney, G.P., et al.: A comparison of similarity measures for use in 2-D-3-D medical image registration. IEEE TMI **17**(4), 586–595 (1998)
5. Heinrich, M.P., et al.: Uncertainty estimates for improved accuracy of registration-based segmentation propagation using discrete optimisation. In: MICCAI Challenge Workshop on Segmentation: Algorithms, Theory and Applications (2013)
6. Heinrich, M.P., Jenkinson, M., Papież, B.W., Brady, S.M., Schnabel, J.A.: Towards realtime multimodal fusion for image-guided interventions using self-similarities. In: Mori, K., Sakuma, I., Sato, Y., Barillot, C., Navab, N. (eds.) MICCAI 2013. LNCS, vol. 8149, pp. 187–194. Springer, Heidelberg (2013). https://doi.org/10.1007/978-3-642-40811-3_24
7. Heinrich, M.P., Papież, B.W., Schnabel, J.A., Handels, H.: Non-parametric discrete registration with convex optimisation. In: Ourselin, S., Modat, M. (eds.) WBIR 2014. LNCS, vol. 8545, pp. 51–61. Springer, Cham (2014). https://doi.org/10.1007/978-3-319-08554-8_6
8. Popuri, K., Cobzas, D., Jägersand, M.: A variational formulation for discrete registration. In: Mori, K., Sakuma, I., Sato, Y., Barillot, C., Navab, N. (eds.) MICCAI 2013. LNCS, vol. 8151, pp. 187–194. Springer, Heidelberg (2013). https://doi.org/10.1007/978-3-642-40760-4_24
9. Heinrich, M.P., et al.: MIND: modality independent neighbourhood descriptor for multi-modal deformable registration. MedIA **16**(7), 1423–1435 (2012)
10. Klein, A., et al.: Evaluation of 14 nonlinear deformation algorithms applied to human brain MRI registration. Neuroimage **46**(3), 786–802 (2009)

11. Shen, D., Davatzikos, C.: HAMMER: hierarchical attribute matching mechanism for elastic registration. IEEE TMI **21**(11), 1421–1439 (2002)
12. Ou, Y., et al.: DRAMMS: deformable registration via attribute matching and mutual-saliency weighting. MedIA **15**(4), 622–639 (2011)
13. Shakeri, M., et al.: Prior-based coregistration and cosegmentation. In: Ourselin, S., Joskowicz, L., Sabuncu, M.R., Unal, G., Wells, W. (eds.) MICCAI 2016. LNCS, vol. 9901, pp. 529–537. Springer, Cham (2016). https://doi.org/10.1007/978-3-319-46723-8_61
14. Wu, G., et al.: Scalable high-performance image registration framework by unsupervised deep feature representations learning. IEEE TBME **63**(7), 1505–1516 (2016)
15. Yang, X., et al.: Quicksilver: fast predictive image registration-a deep learning approach. NeuroImage **158**, 378–396 (2017)
16. Rohé, M.-M., Datar, M., Heimann, T., Sermesant, M., Pennec, X.: SVF-net: learning deformable image registration using shape matching. In: Descoteaux, M., Maier-Hein, L., Franz, A., Jannin, P., Collins, D.L., Duchesne, S. (eds.) MICCAI 2017. LNCS, vol. 10433, pp. 266–274. Springer, Cham (2017). https://doi.org/10.1007/978-3-319-66182-7_31
17. De Brebisson, A., Montana, G.: Deep neural networks for anatomical brain segmentation. In: Proceedings of the IEEE CVPR Workshops, pp. 20–28 (2015)
18. Krizhevsky, A., et al.: ImageNet classification with deep convolutional neural networks. In: NIPS, pp. 1097–1105 (2012)
19. Masci, J., Meier, U., Cireşan, D., Schmidhuber, J.: Stacked convolutional auto-encoders for hierarchical feature extraction. In: Honkela, T., Duch, W., Girolami, M., Kaski, S. (eds.) ICANN 2011. LNCS, vol. 6791, pp. 52–59. Springer, Heidelberg (2011). https://doi.org/10.1007/978-3-642-21735-7_7
20. Ronneberger, O., Fischer, P., Brox, T.: U-net: convolutional networks for biomedical image segmentation. In: Navab, N., Hornegger, J., Wells, W.M., Frangi, A.F. (eds.) MICCAI 2015. LNCS, vol. 9351, pp. 234–241. Springer, Cham (2015). https://doi.org/10.1007/978-3-319-24574-4_28
21. Nyúl, L.G., Udupa, J.K., Zhang, X.: New variants of a method of MRI scale standardization. IEEE TMI **19**(2), 143–50 (2000)
22. Mehta, R., Sivaswamy, J.: M-net: a convolutional neural network for deep brain structure segmentation. In: 2017 IEEE 14th International Symposium on Biomedical Imaging (ISBI 2017), pp. 437–440. IEEE (2017)

Evaluation of Strategies for PET Motion Correction - Manifold Learning vs. Deep Learning

James R. Clough[✉], Daniel R. Balfour, Claudia Prieto, Andrew J. Reader, Paul K. Marsden, and Andrew P. King

School of Bioengineering and Imaging Science, King's College London, London, UK
james.clough@kcl.ac.uk

Abstract. Image quality in abdominal PET is degraded by respiratory motion. In this paper we compare existing data-driven gating methods for motion correction which are based on manifold learning, with a proposed method in which a convolutional neural network learns estimated motion fields in an end-to-end manner, and then uses those estimated motion fields to motion correct the PET frames. We find that this proposed network approach is unable to outperform manifold learning methods in the literature, in terms of the image quality of the motion corrected volumes. We investigate possible explanations for this negative result and discuss the benefits of these unsupervised approaches which remain the state of the art.

Keywords: Motion estimation · Positron emission tomography
Convolutional neural network · Principal component analysis

1 Introduction

Positron emission tomography (PET) imaging is widely used for cancer management and provides information which is vital for diagnosis and monitoring of treatment. In the clinical setting PET is limited by a low signal-to-noise ratio (SNR) because high tracer doses, which increase SNR, also cause radiation exposure and cancer risk to the patient and so the dose is deliberately kept low.

Bodily motion is a further complicating factor which degrades image quality by causing blurring and image artefacts. In particular, respiratory motion is hard to avoid as it is involuntary and many minutes are required to perform a typical PET scan making breath-holding impossible for patients. One way of accounting for organ motion is to simultaneously acquire another imaging modality, such as magnetic resonance (MR) imaging, which can be used to motion correct the PET data. Simultaneous PET-MR scanners [7,12], in which MR can be used to

This work was supported by the Engineering and Physical Sciences Research Council under Grant EP/M009319/1 and by the Wellcome EPSRC Centre for Medical Engineering at Kings College London (WT203148/Z/16/Z).

D. Stoyanov et al. (Eds.): MLCN 2018/DLF 2018/iMIMIC 2018, LNCS 11038, pp. 61–69, 2018.
https://doi.org/10.1007/978-3-030-02628-8_7

motion correct the PET (eg. [2]), are beginning to be used clinically but make up only a small minority of existing PET scanners. Where simultaneous scans are not possible motion modelling using sequential scans can be used for motion correction (eg. [1]). However, motion modelling with sequential scans is limited in its accuracy by the assumption that the breathing patterns during the two scans do not significantly differ.

In principle an attractive solution is to estimate a respiratory signal and to use it to perform motion correction by gating acquired data based on the amplitude of that signal. This signal can be derived from the PET data or from a secondary device measuring, for example, chest position. Data driven signals are promising in that they require no secondary hardware and are directly related to the organ motion of interest, but face the challenge of extracting an accurate signal from low SNR data. A comparison of several data-driven approaches was presented in [14] which found that manifold learning methods such as PCA and Laplacian Eigenmaps performed well as methods of extracting the respiratory signal, with PCA identified as perhaps more stable in noisy conditions.

Recently, convolutional neural networks (CNN) have been shown to be capable of de-noising images taken under low-light conditions or with very short exposure times [4,10]. Such images suffer from Poisson noise, as does PET. The use of CNNs to de-noise, or map low-dose into high-dose images in PET has also been developed recently. In [16] a residual U-net [11] architecture was used to predict full-dose images from 0.5% dose images. In [3] PET-like images were generated from CT data. It may even be possible to de-noise PET data by training a network to perform the inverse-Radon transform and output a high-quality reconstruction from raw sinogram data as is claimed in [17] although the scalability of such an approach remains a significant challenge.

CNNs have also been shown to be capable of performing non-rigid image registration [15]. Such methods can potentially be orders of magnitude faster in their run-time than traditional iterative approaches. Although training the network (i.e. learning the function required to deform each image) is slow, one forward pass (i.e. evaluating that function once to perform the registration) is fast. This approach has proved successful in cases of 2D cardiac MR [15], 3D brain MR [8] and on X-ray images [9]. Notably though, as far as we are aware, such approaches have not been applied to PET imaging, presumably because of the difficulty of dealing with low SNR and non-Gaussian noise. One might then expect that combining these two approaches could allow an appropriate CNN architecture to de-noise a PET frame and estimate the deformation required to transform it to a reference position, which would allow motion correction of a sequence of such frames.

In this paper we attempt to estimate the motion states of PET frames, by training on motion fields acquired from simultaneously acquired MR volumes. We compare a CNN-based approach with a state of the art approach based on manifold learning. We find that, despite our experimentation with various network architectures the CNN approach is unable to outperform the much simpler manifold learning approach.

2   Methods

2.1   Network Architecture

To estimate motion fields directly from time-resolved PET frames we propose a CNN which is illustrated in Fig. 1. The network receives two PET frames (see Fig. 2 for examples) as its input - the 3D volume from a reference respiratory position R, which is a fully exhaled position, and the 3D volume in question, V_t. In our experiments these volumes are both of size $48 \times 176 \times 256$ in the anterior-posterior $\times$ head-foot $\times$ left-right directions. The desired output is the set of three-dimensional motion fields, M_t which represent the deformation of the underlying anatomy from the position in V_t to the position in R, which can then be used to transform V_t into the reference motion state. The ground truth motion fields are such that $M_t(V_t) \approx R$, where $M_t(V_t)$ denotes the result of applying the transformation M_t to the volume V_t.

The output of the network is a $48 \times 176 \times 256 \times 3$ tensor, representing the required voxelwise deformation in the x, y, z directions. As a loss function we use the mean square difference between the three components of the predicted motion field vectors and the ground truth components which simply corresponds to the mean square displacement between the predicted and ground truth vectors.

2.2   Training Details

Our method was implemented in Keras[1]. The network was trained with the Adam optimiser, with a learning rate of 0.001, and with Dropout regularisation in the two convolutional layers in the lowest resolution layer of the U-net. We used a batch size of 4, the maximum allowed by our GPU memory. In all cases, the results for one subject are acquired by training the network on all PET frames from all other subjects.

3   Experiments

3.1   Synthetic Dataset

We conducted our experiments on a highly-realistic synthetic dataset. The data consist of real MR acquisitions which are then used to create synthetic PET data, giving us a paired PET-MR dataset. The PET simulations were intended to mimic a typical ^{18}F-fluorodeoxyglucose (FDG) scan. Cardiac-gated abdominal MR scans were performed on 10 healthy volunteers, with both a high-resolution exhale breath-hold volume, and sequences of 35 low-resolution dynamic volumes acquired for each of three breathing modes, 'deep breathing', 'normal breathing' and 'fast breathing' making 105 acquired low-resolution dynamic MR volumes for each volunteer.

---

[1] https://keras.io/.

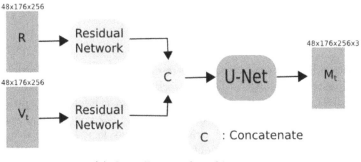

(a) Overall network architecture

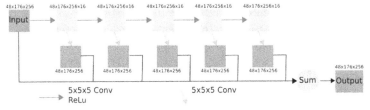

(b) Residual network architecture

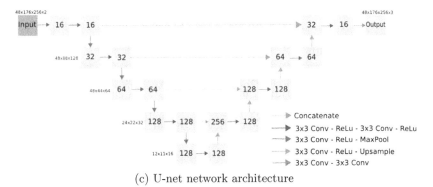

(c) U-net network architecture

Fig. 1. Diagram of the neural network architecture used in these experiments. Both of the input volumes are first passed through a shared residual de-noising layer, before being concatenated and passed into a U-net like architecture to incorporate both local and global information into the final motion estimation.

The high-resolution volumes were segmented into anatomical regions relevant to PET emission and attenuation to create attenuation maps and FDG emission maps for each volunteer. These FDG maps were then augmented by adding artificial lesions (either one or two spherical lesions in the lungs and/or liver, of sizes between 10 mm and 20 mm in diameter) such that each volunteer had ten emission maps (one unmodified, four with one added lesion, and five with two added lesions).

Fig. 2. Examples of typical simulated PET frames from four volunteers. The images shown are coronal sections chosen to make the lesions easily visible.

Motion fields were extracted by performing a non-rigid registration from the high-resolution breath-hold to the low-resolution dynamic volumes. The simulated PET was then created by using the calculated motion fields to warp these attenuation and emission maps, from which PET sinograms were simulated and then time-resolved frames reconstructed using the ordered-subsets expectation maximisation (OSEM) reconstruction algorithm [6]. The simulations include random coincidences and scatter, with each simulated scan having a total of 50 million simulated counts and an additional 25 million random coincidences.

In total this gave us 10 volunteers each with 10 artificial lesion placements, and motion states from 3 breathing modes each with 35 acquired volumes giving a total of $10 \times 10 \times 35 \times 3 = 10500$ simulated PET frames.

Finally, we also simulate PET acquisitions using no motion fields, producing a simulation of a theoretical acquisition in which there was no respiratory motion. This provides us with a best achievable performance for motion correction.

3.2 Comparison Method: Data-Driven Gating

To evaluate the CNN-based motion correction approach we compare it to the unsupervised PCA-based method introduced in [13]. As implemented here, this method involves taking the Freeman-Tukey [5] transformation of the PET frames and then taking the first component of the PCA of this data (which we find always corresponds to respiratory motion) as a gating signal. The 35 PET frames for each sequence are then grouped into 5 gates using this gating signal, the data in each group aggregated, and the resulting volumes then registered to a target gate. The data from these groups are then aggregated to create the final motion corrected volume.

3.3 Assessment of Corrected Volume Quality

We quantitatively assessed the quality of the motion corrected volumes using the peak standardised uptake value (SUV) in the region of interest (ROI) of the lesion(s). The SUV for a voxel was found by taking a small region of interest (the voxel in question and its 6 adjacent voxels) and taking the mean intensity value across this group. The voxel within the ROI of the lesions with the highest such mean value determines the peak SUV value in that region. The lesion's ROI was defined by the lesion's position in the original segmentations used to

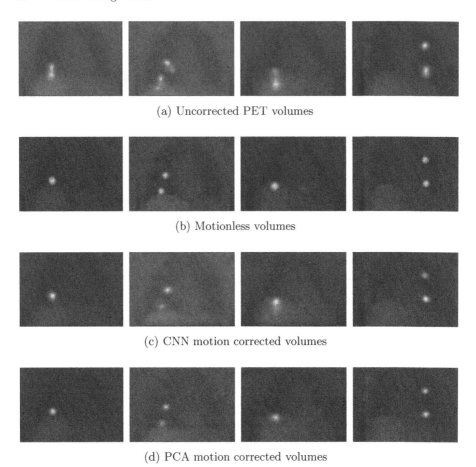

(a) Uncorrected PET volumes

(b) Motionless volumes

(c) CNN motion corrected volumes

(d) PCA motion corrected volumes

Fig. 3. Example of uncorrected volumes (top row), motionless volumes (second row), and motion corrected volumes with the CNN method (third row) and with the PCA method (fourth row), from four of the volunteers in our dataset.

create the simulated PET, which effectively represents a ground-truth position for the lesion. We use the peak SUV calculated in this way to compute our final evaluation metric which is a percentage SUV recovery. The peak SUV values for the motion correction methods assessed here are expressed as a percentage of the motionless peak SUV value. For the CNN motion correction method, we used a cross-validation scheme in which the CNN was trained on the other nine volunteers and then tested on the left-out volunteer. As is clear from Table 1, while both methods of motion correction improve upon the raw, uncorrected volumes, the PCA method outperforms the CNN method on all ten volunteers. Examples of motion corrected volumes are shown in Fig. 3.

Table 1. Percentage of peak SUV recovered using motion correction, with the gold-standard motionless volumes SUV values set to be 100% for each volunteer. Shown here are the mean and standard deviation of the peak SUV fraction over the nine artificial lesion placements, excluding the tenth case where no lesion was present.

Volunteer	Uncorrected	PCA corrected	CNN corrected
1	45.5% ± 11.0	68.6% ± 3.3	57.5% ± 7.4
2	43.4% ± 11.3	71.1% ± 2.5	57.9% ± 4.0
3	30.8% ± 3.6	60.4% ± 4.0	37.4% ± 6.3
4	32.1% ± 1.4	62.0% ± 7.7	38.1% ± 4.2
5	32.0% ± 5.9	71.9% ± 14.2	43.4% ± 5.8
6	31.3% ± 6.4	70.8% ± 7.4	39.4% ± 5.0
7	61.1% ± 8.9	77.9% ± 6.2	69.3% ± 5.1
8	58.6% ± 11.7	76.1% ± 5.6	65.0% ± 7.7
9	66.9% ± 4.4	75.3% ± 4.9	66.7% ± 5.6
10	62.5% ± 8.7	79.0% ± 4.7	70.9% ± 8.1

4   Discussion and Conclusions

Why might unsupervised methods be more powerful or more appropriate solutions for motion correction in PET than deep CNNs? Although breathing patterns between patients vary significantly, the breathing pattern for one patient over a short amount of time is often well modelled by a low-dimensional manifold [2]. This is especially true when the relevant signal in the image is highly concentrated in space, as is the case here where small lesions with high levels of FDG emission are the most important structures for motion correction. If the lesion repetitively traces out a path during respiration, and it is significantly brighter than the rest of the volume, then this signal is likely to be easily picked up by manifold learning techniques, as has been demonstrated here. More complicated organ motion which cannot be inferred from the lesion motion will not be picked up by manifold learning approaches, but importantly this kind of motion outside of the lesions will not affect the clinically relevant measurement of image quality such as the peak SUV as used here, or alternative measures such as lesion size or position.

Simple manifold learning methods may be more restrictive than CNN-based methods but in cases where the training data are very noisy, and the signal being estimated is low-dimensional, these restrictions seem to be beneficial. We note that as well as the CNN architecture described here we attempted to use several modifications which proved not to help the final image quality, including changing the sizes of the convolution kernels, numbers of layers and feature maps, and estimating joint motion fields from temporally neighbouring frames to make use of temporal correlations. We also found that, with sufficiently long training times, the CNN was able to accurately fit the training set motion fields suggesting that the problem on the test set is one of generalising to unseen subject's anatomies and breathing patterns, although further work is required to understand exactly to what extent these differences limit the final motion-corrected image quality.

While our experiments cannot demonstrate that all CNN-based methods for motion correcting PET data will struggle, they do suggest that at the very least, when the training set is relatively small, it is very challenging to construct a CNN motion correction method for PET which approaches the performance of the state-of-the-art manifold learning methods.

Acknowledgments. We would like to thank nVidia for kindly donating the Quadro P6000 GPU used in this research.

References

1. Balfour, D.R., et al.: Respiratory motion correction of PET using MR-constrained PET-PET registration. Biomed. Eng. Online **14**(1), 85 (2015)
2. Baumgartner, C.F., et al.: High-resolution dynamic MR imaging of the thorax for respiratory motion correction of PET using groupwise manifold alignment. Med. Image Anal. **18**(7), 939–952 (2014)
3. Ben-Cohen, A., Klang, E., Raskin, S.P., Amitai, M.M., Greenspan, H.: Virtual PET images from CT data using deep convolutional networks: initial results. In: Tsaftaris, S.A., Gooya, A., Frangi, A.F., Prince, J.L. (eds.) SASHIMI 2017. LNCS, vol. 10557, pp. 49–57. Springer, Cham (2017). https://doi.org/10.1007/978-3-319-68127-6_6
4. Chen, C., et al.: Learning to see in the dark. arXiv preprint arXiv:1805.01934 (2018)
5. Freeman, M.F., Tukey, J.W.: Transformations related to the angular and the square root. Ann. Math. Stat. **21**(4), 607–611 (1950)
6. Hudson, H.M., Larkin, R.S.: Ordered subsets of projection data. IEEE Trans. Med. Imaging **13**(4), 601–609 (1994)
7. Judenhofer, M.S.: Simultaneous PET-MRI: a new approach for functional and morphological imaging. Nat. Med. **14**(4), 459–465 (2008)
8. Li, H., Fan, Y.: Non-rigid image registration using self-supervised fully convolutional networks without training data. arXiv preprint arXiv:1801.04012 (2018)
9. Miao, S.: A CNN regression approach for real-time 2D/3D registration. IEEE Trans. Med. Imaging **35**(5), 1352–1363 (2016)
10. Remez, T., et al.: Deep convolutional denoising of low-light images (2017). http://arxiv.org/abs/1701.01687
11. Ronneberger, O., Fischer, P., Brox, T.: U-net: convolutional networks for biomedical image segmentation. In: Navab, N., Hornegger, J., Wells, W.M., Frangi, A.F. (eds.) MICCAI 2015. LNCS, vol. 9351, pp. 234–241. Springer, Cham (2015). https://doi.org/10.1007/978-3-319-24574-4_28
12. Shao, Y., et al.: Simultaneous PET and MR imaging. Phys. Med. Biol. **42**(10), 1965 (1997)
13. Thielemans, K., et al.: Device-less gating for PET/CT using PCA. In: IEEE Nuclear Science Symposium Conference Record, pp. 3904–3910 (2011)
14. Thielemans, K., et al.: Comparison of different methods for data-driven respiratory gating of PET data. In: IEEE Nuclear Science Symposium Conference Record, pp. 3–6 (2013)

15. de Vos, B.D., Berendsen, F.F., Viergever, M.A., Staring, M., Išgum, I.: End-to-end unsupervised deformable image registration with a convolutional neural network. In: Cardoso, M.J., et al. (eds.) DLMIA/ML-CDS -2017. LNCS, vol. 10553, pp. 204–212. Springer, Cham (2017). https://doi.org/10.1007/978-3-319-67558-9_24
16. Xu, J., et al.: 200x low-dose pet reconstruction using deep learning. arXiv preprint arXiv:1712.04119 (2017)
17. Zhu, B., et al.: Image reconstruction by domain transform manifold learning. Nat. Publ. Group **555**(7697), 487–492 (2017)

Exploring Adversarial Examples
Patterns of One-Pixel Attacks

David Kügler[1]([✉]), Alexander Distergoft[1], Arjan Kuijper[2],
and Anirban Mukhopadhyay[1]

[1] Interactive Graphics Systems Group, Technische Universität Darmstadt,
Darmstadt, Germany
david.kuegler@gris.tu-darmstadt.de
[2] Fraunhofer IGD, Darmstadt, Germany

Abstract. Failure cases of black-box deep learning, e.g. adversarial examples, might have severe consequences in healthcare. Yet such failures are mostly studied in the context of real-world images with calibrated attacks. To demystify the adversarial examples, rigorous studies need to be designed. Unfortunately, complexity of the medical images hinders such study design directly from the medical images. We hypothesize that adversarial examples might result from the incorrect mapping of image space to the low dimensional generation manifold by deep networks. To test the hypothesis, we simplify a complex medical problem namely pose estimation of surgical tools into its barest form. An analytical decision boundary and exhaustive search of the one-pixel attack across multiple image dimensions let us localize the regions of frequent successful one-pixel attacks at the image space.

Keywords: CNN · Adversarial examples · One-pixel attack
Deep Learning Fails

1 Introduction

End-to-end Deep Learning pipelines (image in, classification out) have achieved significant success in Medical Image Computing (MIC) across multiple scenarios, even stretching to Computer-Aided Interventions (CAI) [6]. This success in comparison to traditional methods (including based on learning) has hurried an AI-summer in Healthcare seen in the prevalence of Deep Learning-publications in the MICCAI community. Political authorities have recognized this shift towards Deep Learning-based methods and are taking action. In the United States, the FDA has embraced this change by approving AI devices for diabetic retinopathy detection [1] and is currently in the discussion towards easing the approval process for AI-based medical software [7]. The European Union, on the other hand, has introduced the General Data Protection Regulation, which necessitates the right to explanation of any decisions taken by a computerized system.

Since researchers struggle to explain decisions by Deep Learning models, the underlying function is yet a Black Box in practice. Its analysis is hindered by the

© Springer Nature Switzerland AG 2018
D. Stoyanov et al. (Eds.): MLCN 2018/DLF 2018/iMIMIC 2018, LNCS 11038, pp. 70–78, 2018.
https://doi.org/10.1007/978-3-030-02628-8_8

difficulty of deriving and understanding the decision boundary. In fact, recent studies [2,5,8,9,11] have shown that these Deep Learning models are vulnerable to adversarial examples – these are images which cause incorrect classifications despite either models predicting with high certainty or being clear classifications to humans. Adversarial examples are not understood as a consequence of the black-box-characteristic. They can even be as simple as only changing a single pixel leading to different classification results (one-pixel-attacks). For medical applications, the exploitation of this vulnerability is thoroughly analyzed by Finlayson et al. [3]. However, this vulnerability is largely ignored across evaluations of Deep Learning in MIC and CAI.

Though impressive, the general example attacks shown by Finlayson et al. [3] address the traditional image-in-diagnosis-out setting. However, real medical images and annotations are complex further obscuring the situation and adding to the mystery – for example complex image structure, confounding situations (device vendors, acquisition parameters), non-conformity between radiologists, multi-class and multi-label decisions. Without disentangling these factors it is impossible to understand adversarial examples, which stem from Deep Learning only. Simplifications of MIC scenarios are needed to draw systematic conclusions regarding adversarial examples. For example: to consider segmentation masks rather than the real-world images (binary instead of continuous pixel values) or to define decision boundaries in a closed form (which is not available in MIC).

In this paper, we provide the first systematic analysis of one-pixel-attacks on convolutional neural networks (CNNs) in a simplified CAI application and provide a first intuition of patterns. With inherent limits on the knowledge and processable size of both the likely image space and a complete description of the decision boundary, it will be impossible to analyze adversarial attacks due to the complexity of CNNs – even here simplifications are needed. To break the problem down to its core, we (a) simplify the range of images (image space), (b) train multiple classifiers and (c) search exhaustively for one-pixel adversarial examples. We consider the problem of instrument pose estimation studied by Kügler et al. [6], who have ignored adversarial attacks, where the orientation of instruments is to be determined. To gain control over the image and annotation complexity, we define a continuous generation manifold with a perfectly defined binary decision boundary. From individual manifold coordinates, we generate binary images at various dimensions with the instrument being simplified to a line for different levels of discretization. We define all images that can be generated through this pipeline as "possible images" and train multiple simple classifiers based on convolutional neural networks with "ALL" these uniquely possible images. Finally, we exhaustively search the space of all single pixel-flip adversarial candidates, identify successful attacks and localize the regions of frequent successful one-pixel-attacks. The most surprisingly, the overwhelming majority of attacks are localized at a distance of the instrument, which implies the one-pixel-flip *did not change the information of the image*.

2   Related Work

Goodfellow *et al.* [5] demonstrate that standard image models exhibit a strange phenomenom: most randomly chosen images from a data distribution are correctly classified and yet are close to visually similar images that are incorrectly classified. By adding some certain kind of perturbation to an image this behaviour can be reproduced on most CNNs. A hypothesis on that behaviour is that neural network classifiers are too linear in various regions of the input space (Fig. 1).

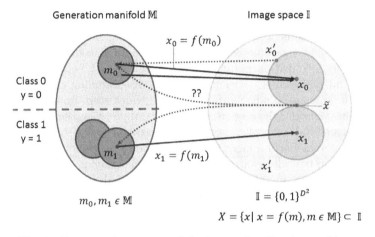

Fig. 1. Conceptual summary of the image classification problem.

Su *et al.* [10] specialize on so called 'one pixel attacks'. They show that even by adding a perturbation with the size a single pixel to an image, the output of deep learning networks can easily be altered.

Gilmer *et al.* [4] try to get an insight on adversarial examples by training different neural networks on a synthetic dataset of two concentric spheres with different radii. The idea is to classify whether a point belongs to one or the other sphere. Even though the data manifold as well as the theoretical max margin boundary are clearly defined and enough input is provided for networks to train on, adversarials can still be found near correctly classified points.

3   Methods

We generate a custom dataset to study adversarial examples for image classification. Using a exhaustive search, all one-pixel-flip candidate images are tested for misclassification.

Analyzing the space $\mathbb{I}$ of images, we differentiate between images belonging to an application ($\mathbb{I}_{\tilde{U}}^* \subset \mathbb{I}$, often termed "natural images") and images holding

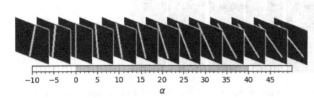

Fig. 2. Generated images from a subset. The bar below indicates the decision boundary. For the green range $y = 1$, $y = 0$ otherwise. Generated images are colorized for better contrast. (Color figure online)

Table 1. Number of unique images for different combinations of image dimensions and rotation stepsizes

Dimensions	0.5°	1.0°	2.0°
16 × 16	394	280	170
32 × 32	2524	1394	716
48 × 48	5244	2790	1436
64 × 64	8116	4228	2158
80 × 80	10998	5676	2880

no information on the application. By introducing a generation manifold $\mathbb{M}$ – a higher-level parameter space describing all images possible for our application – we are able to describe all features of the image relevant to our application.

For a systematic analysis of adversarial examples, we need a closed-form representation of the decision boundary. Defining the decision boundary dependent on the generation manifold allows us to determine the distance of an image to the decision boundary in terms of the manifolds coordinates. The generation function $f : \mathbb{M} \to \mathbb{I}_U^*$ maps from the generation manifold to the image space $\mathbb{I}_U^*$ restricted to images that can be generated by f. Since all generations from a region around $m_i \in \mathbb{M}$ lead to identical images in $\mathbb{I}$, those cases are in-differentiable, i.e. f is non-injective. In addition, since not all images from $\mathbb{I}$ can be created by this function, some images $x \in \mathbb{I}$ do not have a corresponding coordinate m in $\mathbb{M}$. For images that cannot be directly generated through the generation function, but that are largely similar to images directly generated (e.g. one-pixel attacks x_i' of image x_i), the association to the corresponding m_i is implicit (dotted line). Some images $\tilde{x}$ will have the property of being equally similar to two different m even belonging to different classes. For these ambiguous images, no association to a m or even a class can be found.

3.1   Dataset

With our chosen generation manifold $\mathbb{M}$ as (L, α), our generation function maps to images of lines of varying length L and rotation α (see Fig. 2). Lines are centered on the image center, leading to a scenario where images always differ in at least 2 pixels because of symmetry. A line-like structure is being used, to keep the generation manifold as simple as possible. Finally, we define a simple decision boundary classifying images into 2 categories, where $y = 1$ for α in the range from 0° to 40°, and $y = 0$ for all other cases. The chosen range for α is arbitrary.

We generate 15 complete sets of all unique binary images $X = \mathbb{I}_U^*$ for 15 different generation functions f by different manifold discretization (only allowing α in steps of 0.5°, 1.0° and 2.0°) and image dimensions $D \times D$ (16 × 16, 32 × 32,

Fig. 3. Left to right: image from dataset for $0°$; 2 confirmed exemplary adversarial candidates.

48×48, 64×64 and 80×80). The length L is varied between 12 Pixel and $D - 2$, where D is the width and the height of the image. This procedure leads to varying numbers of unique images ($|X|$) as described in ţ.

3.2   Training

We train 5 models for each of the 15 synthetic combinations (Table 1), resulting in a total of 75 trained networks. Training repitions were performed to average out the stochastic properties of the training process.

Since we are looking at off-manifold images rather than the generalization error of the networks, there is no need for a testing set. Moreover, since we use every single image in a subset (i.e. every possible image) for training, our models are not restricted by the choice of training data. All models feature the same architecture only differing in the dimensions of the input images. We design a simple network architecture with three layers: First, two 2D convolutional layers of size 3×3 with 32 channels and ReLU activation each followed by max pooling (stride 2) process the input. On this, a fully connected layer with 128 units and ReLU activation is applied, followed by an output layer with 2 units and a Sigmoid activation. We regulize by dropout ($p = 0.25$) just before the fully connected layer.

For optimization, we use the Adam optimizer with recommended parameters ($\beta_1 = 0.9$ and $\beta_2 = 0.999$) and a learning rate of 0.001. Finally, binary crossentropy is used as the loss function. We achieved an accuracy of 1.0 with all models indicating perfect convergence on the training dataset.

3.3   Adversarial Data Creation

The goal of creating adversarial data is to identify whether our trained networks can be fooled into predicting the wrong output y for a given image x. By flipping one pixel at a time, we perform an exhaustive search of all combinations of all images in X (see Fig. 3 for examples). The total number of adversarial candidates $N_{adv} = ND^2$, where $N = |X|$ and D being the width and length of the image. Unlike Su *et al.* [10], we brute-force our way to a *complete list of all possible adversarial examples* instead of finding single instances by optimization, which did not work for binary images.

4   Results

By testing the classification of all adversarial candidates, we found *all networks* to be vulnerable to one-pixel-adversarial attacks. All experiments were performed on 5 networks, so all values are averages over 5 networks.

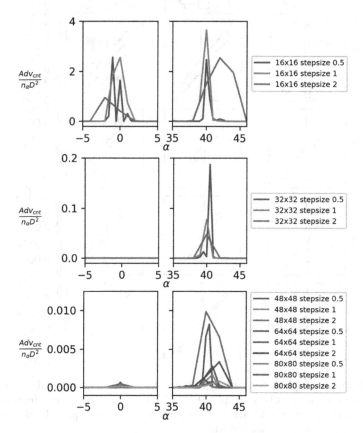

Fig. 4. Comparison of average adversarial likelihood depending on the distance to decision boundary. Plots show the areas around $\alpha = 0°$ and $\alpha = 40°$.

We evaluated the relative number of adversarial examples w.r.t. the adversarial candidates. We also determined this ratio $ADV_{cnt}/(ND^2)$ of actual adversarial examples to adversarial candidates dependent on the angle rotation α ($Adv_{cnt}/(n_\alpha D^2)$, see Fig. 4) and on the pixel-position in the image (see Fig. 5). These ratios can also be interpreted as experimentally determined average likelihood of an image being adversarial given the specific conditions.

Figure 4 shows the distribution of the relationship of misclassifications at a particular angle. Adversarial examples were only found around the decision boundary ($\alpha \approx 0°$ or $\alpha \approx 40°$. With an increase of the image dimension D the likelihood to find adversarial examples *decreases*.

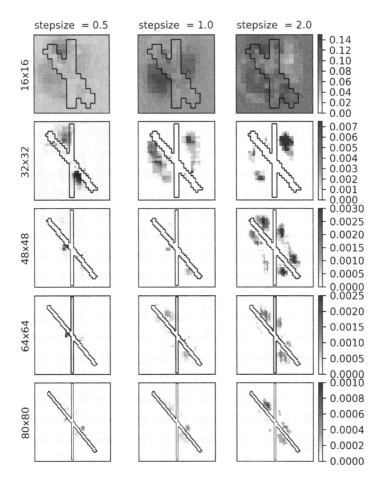

Fig. 5. Heatmaps display the spatial likelihood for a flip to cause a misclassification, i.e. leading to an adversarial example; the "X" represent edges of two images on the decision boundary, i.e. $\alpha = 0°$ and $\alpha = 40°$; interestingly, high probabilities are often found in regions removed from the edges

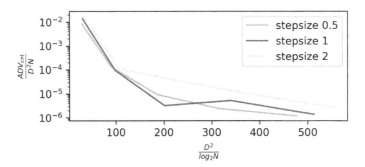

Fig. 6. Information or compression analysis

We created heatmaps, shown in Fig. 5, to display the spatial likelihoods for a flip to cause a missclassification. The "X" represented the edges of lines from two images, which belonged to the values of α on the decision boundary, i.e. $\alpha = 0°$ and $\alpha = 40°$. Surprisingly, the highest likelihoods to cause an image to be confirmed as adversarial were not situated at positions, where the discrimination between classes got harder from an information perspective, but were removed from the edges, i.e. the overwhelming majority of images was *not ambiguous*. Increasing the image dimension led to more pronounced patterns and better overall robustness. This can also be seen in Fig. 6.

Finally, we analyzed whether the relationship between the relative number of adversarial examples and the "possible redundancy" of the reconstructable manifold information in image space exits. The latter is the ratio of information contained in the reconstructable manifold and the image. Higher "redundancy" seems to indicated a strong relationship to increased robustness to adversarial attacks.

Unlike other studies, these results were obtained for networks trained on *"ALL" possible images* that can be generated from the generation manifold and achieved an accuracy of 1.0.

5   Discussion and Conclusion

This paper provides a systematic evaluation of one pixel adversarial attacks on convolutional neural networks. By leveraging a simple toy CAI scenario against a simple yet perfect (accuracy 1.0) convolutional neural network, we find one-pixel-adversarial-candidates with an astonishing regularity. These candidates are deliberately placed close to but off the manifold training images are drawn from. In particular, we identify the vulnerable regions to be close to the decision boundary and not explainable by loss of informations caused by the introduction of the "attacking pixel".

In Future Work, we will generalize these observations to toy examples derived from other scenarios, increase the depth of the neural network and investigate causes for adversarial examples.

The systematic analysis of adversarial examples presented in this paper initiates a much needed process of understanding adversarial examples in medical images.

References

1. Berkrot, B.: U.S. FDA approves AI device to detect diabetic eye disease, 11 April 2018. https://www.reuters.com/article/us-fda-ai-approval/u-s-fda-approves-ai-device-to-detect-diabetic-eye-disease-idUSKBN1HI2LC
2. Eykholt, K., et al.: Robust physical-world attacks on deep learning models (2017). http://arxiv.org/pdf/1707.08945
3. Finlayson, S.G., Chung, H.W., Kohane, I.S., Beam, A.L.: Adversarial attacks against medical deep learning systems (2018). http://arxiv.org/pdf/1804.05296

4. Gilmer, J., et al.: Adversarial spheres (2018). http://arxiv.org/pdf/1801.02774
5. Goodfellow, I.J., Shlens, J., Szegedy, C.: Explaining and harnessing adversarial examples (2014). http://arxiv.org/pdf/1412.6572
6. Kügler, D., Stefanov, A., Mukhopadhyay, A.: i3PosNet: instrument pose estimation from X-Ray (2018). http://arxiv.org/pdf/1802.09575
7. Walter, M.: FDA reclassification proposal could ease approval process for CAD software, 01 June 2018. https://www.healthimaging.com/topics/healthcare-economics/fda-reclassification-proposal-could-ease-approval-process-cad-software
8. Nguyen, A., Yosinski, J., Clune, J.: Deep neural networks are easily fooled: high confidence predictions for unrecognizable images (2014). http://arxiv.org/pdf/1412.1897
9. Rauber, J., Brendel, W., Bethge, M.: Foolbox: a python toolbox to benchmark the robustness of machine learning models (2017). http://arxiv.org/pdf/1707.04131
10. Su, J., Vargas, D.V., Kouichi, S.: One pixel attack for fooling deep neural networks (2017). http://arxiv.org/pdf/1710.08864
11. Szegedy, C., et al.: Intriguing properties of neural networks (2013). http://arxiv.org/pdf/1312.6199

Shortcomings of Ventricle Segmentation Using Deep Convolutional Networks

Muhan Shao[1(✉)], Shuo Han[2,3], Aaron Carass[1,4], Xiang Li[1], Ari M. Blitz[5], Jerry L. Prince[1,4], and Lotta M. Ellingsen[1,6]

[1] Department of Electrical and Computer Engineering,
The Johns Hopkins University, Baltimore, MD 21218, USA
muhan@jhu.edu
[2] Department of Biomedical Engineering,
The Johns Hopkins University School of Medicine, Baltimore, MD 21205, USA
[3] Laboratory of Behavioral Neuroscience, National Institute on Aging,
National Institutes of Health, Baltimore, MD 20892, USA
[4] Department of Computer Science, The Johns Hopkins University,
Baltimore, MD 21218, USA
[5] Department of Radiology and Radiological Science, The Johns Hopkins University,
Baltimore, MD 21287, USA
[6] Department of Electrical and Computer Engineering, University of Iceland,
Reykjavik, Iceland

Abstract. Normal Pressure Hydrocephalus (NPH) is a brain disorder that can present with ventriculomegaly and dementia-like symptoms, which often can be reversed through surgery. Having accurate segmentation of the ventricular system into its sub-compartments from magnetic resonance images (MRI) would be beneficial to better characterize the condition of NPH patients. Previous segmentation algorithms need long processing time and often fail to accurately segment severely enlarged ventricles in NPH patients. Recently, deep convolutional neural network (CNN) methods have been reported to have fast and accurate performance on medical image segmentation tasks. In this paper, we present a 3D U-net CNN-based network to segment the ventricular system in MRI. We trained three networks on different data sets and compared their performances. The networks trained on healthy controls (HC) failed in patients with NPH pathology, even in patients with normal appearing ventricles. The network trained on images from HC and NPH patients provided superior performance against state-of-the-art methods when evaluated on images from both data sets.

Keywords: MRI · Hydrocephalus · Segmentation · CNN

1 Introduction

The ventricular system of the human brain is composed of four interconnected cavities: the left and right lateral, the third and the fourth ventricles. Each ventricle contains choroid plexus, a network of ependymal cells producing cerebrospinal

© Springer Nature Switzerland AG 2018
D. Stoyanov et al. (Eds.): MLCN 2018/DLF 2018/iMIMIC 2018, LNCS 11038, pp. 79–86, 2018.
https://doi.org/10.1007/978-3-030-02628-8_9

fluid (CSF). Normal pressure hydrocephalus (NPH) is a brain disorder usually caused by disruption of CSF flow but with normal CSF pressure. The ventricles expand and press against the brain tissue nearby, which can lead to the distortion of the brain shape and eventually cause brain damage. NPH is characterized by gait unsteadiness, urinary incontinence, and dementia [1]. However, unlike most forms of dementia, the symptoms in NPH are potentially reversible to a certain extent on properly selected patients. Diversion of CSF through shunt surgery has been reported to improve the symptoms of NPH [10]. However, it remains a challenge to identify NPH patients who respond to treatment, and differentiate NPH from other neurodegenerative disorders, such as Alzheimer's disease [11].

Currently, NPH is diagnosed based on characteristic clinical symptoms and brain imaging [11]. The ventricular dilation in NPH can be observed through magnetic resonance (MR) images. Examples of T1-weighted (T1w) Magnetically Prepared Rapidly Acquired Gradient Echo (MPRAGE) images of NPH patients are shown in Fig. 3(a). Disproportionate dilation of components of the ventricular system in NPH is relative to the specific point of CSF disruption, which could have an impact on the diagnosis [11]. Therefore, accurate segmentation of the ventricular system into its four cavities could help characterize the pathophysiology and potentially lead to better surgical planning of NPH patients.

Previously published segmentation methods include the popular FreeSurfer [6] method and many multi-atlas segmentation methods [15,20]. However, these methods require long processing times (several hours) and often fail to capture the boundary of the greatly enlarged ventricles in NPH patients. A recently developed segmentation algorithm, RUDOLPH [3,5], is a combined patch-based and multi-atlas segmentation method designed for subjects with ventriculomegaly. Although this method is robust in ventricular parcellation, it also has a long runtime. In recent years, various methods based on deep convolutional neural networks (CNN) have been proposed to tackle neuroimage segmentation [2,12]. The U-Net [16] is one of the most well-known CNN architectures in medical image analysis. The skip connections between contracting and expanding paths in the U-Net improve the network performance.

In this paper, we present a 3D U-Net method for segmenting the ventricular system. We trained three networks on images from two data sets, two comprising healthy controls (HC) and the other a mix of HC and NPH patients, and show the difference of their performances. The first network was trained on 13 HC and performed well when evaluated on subjects from the same data set. However, it performed poorly on the NPH data set, even on images with normal sized ventricles. The second network was trained on 38 HC, including elderly subjects with enlarged ventricles, and performed even worse than the first network when evaluated on NPH data set. The third network was trained on a mixture of 13 HC and 25 NPH images and provided dramatically improved results on both data sets, demonstrating the importance of training data selection.

2   Methods

2.1   Data and Preprocessing

We evaluated our segmentation network using 3D brain MR images from two data sets. The first one comprised 38 T1w MR images from Neuromorphometrics Inc (NMM)[1]. Each image was manually delineated by experts into 138 brain structures. For our purposes, we converted the 139 labels (138 brain structure labels and 1 background label) into five: left and right lateral ventricles, third ventricle, fourth ventricle, and a catch-all background label. The inferior lateral ventricle label was included with the corresponding lateral ventricle label. The T1w MR images were sorted by the volume of the ventricular system and 13 images were used as training data for the first and third network, covering the entire spectrum of ventricle sizes in the data set. All 38 images were used as training data for the second network.

The second data set was from our NPH database comprising 95 NPH patients with a wide range of ventriculomegaly. They were acquired on a 3T (Siemens Corporation, Germany) scanner with T1w MPRAGE with TR = 10.3 ms, TE = 6 ms, and $0.82 \times 0.82 \times 1.17\,\text{mm}^3$ voxel size. We manually delineated the ventricular system in all 95 NPH patients from our database into our five labels. A total of 25 NPH images, ranging from mild to severe cases, were chosen as our training data for the second network.

The images from the two data sets were run through a preprocessing pipeline, including N4 bias correction [18], rigid registration to MNI 152 atlas space [7], and skull stripping [17].

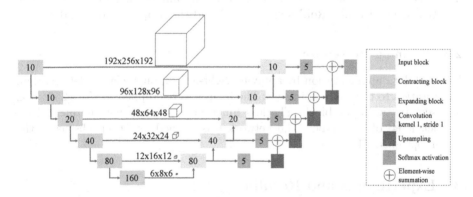

Fig. 1. Architecture of the ventricle-segmentation network. The numbers in the contracting and expanding blocks indicate the output number of features. The shape of the tensor is denoted next to the box in each resolution scale.

---

[1] http://www.neuromorphometrics.com/.

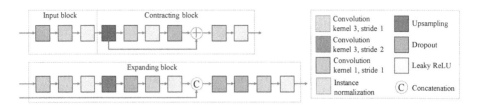

Fig. 2. Architecture of the input, contracting, and expanding blocks used in the segmentation network.

2.2 Ventricle Segmentation Network

A 3D U-Net [13] was modified to segment the left and right lateral ventricles, and the third and fourth ventricles. In this network (Fig. 1), a series of contracting blocks extract image features from local to global context and a series of expanding blocks, with shortcut to contracting blocks, act as "learnable" upsampling interpolation to restore the feature map resolution (Fig. 2). Using learned features, the projection convolution connected to each expanding block (Fig. 1) along with the softmax operation further classify the voxels into five labels including the four ventricles and the background.

The contracting block is similar to the building block for increasing dimensions of the pre-activation ResNet [9], since the shortcut within a block can make the optimization easier and increase accuracy [8]. In contrast to ResNet, however, the identity mapping and the residue encoding paths share the first convolution in this design to reduce overfitting. Instance normalization [19] was used since it is invariant to mean and covariant shift of image intensities. The negative slope of Leaky ReLUs [22] was 0.1 and the dropout rate was 0.2.

2.3 Training Procedure

We used data augmentation by applying right-left flipping, elastic deformation, and rotation to the training images. The images were cropped to $192 \times 256 \times 192$ and sent to the input block. The loss function was one minus the mean Dice coefficient [4] of each label. The network was trained for 50 epochs using the Adam optimizer [14].

3 Experiments and Results

We trained three networks, VenSeg1 using 13 T1w MR images from NMM, VenSeg2 using 38 T1w MR images from NMM, and VenSeg3 using 38 T1w MR images including the same 13 in VenSeg1 and 25 from our NPH cohort. The 95 MR images (25 from NMM and 70 from NPH) formed the testing data set. We only evaluated the performance of VenSeg2 on the 70 NPH testing images.

The 25 testing images from NMM data set were processed by VenSeg1, VenSeg3, and three state-of-the-art brain segmentation methods: FreeSurfer 6.0,

Joint label fusion (JLF) [20] and RUDOLPH [3]. The 70 testing images from the NPH cohort were processed by all the six segmentation methods. We provided FreeSurfer with skull-stripped data to speed up the process and turned on the -bigventricles switch for NPH subjects to handle the enlarged ventricles.

Visual comparisons of the five methods (excluding VenSeg2) on one NMM image and three NPH images are shown in Fig. 3. The VenSeg1 network provided accurate segmentation on the NMM image (Fig. 3(a), subject #1). However, it yielded erroneous segmentations on MR images of NPH patients. A truly surprising failure of VenSeg1 is subject #2; Subject #2 has a similar shape and volume to subject #1 from the NMM cohort (129 ml for subject #2 and 132 ml for subject #1) and yet VenSeg1 failed to capture the boundary of the lateral ventricles and mislabeled portions of the right ventricle as left. Subject #3 in Fig. 3 shows an NPH patient with mild pathology, however VenSeg1 incorrectly labeled some cortex as the 4th ventricle (yellow arrow in Fig. 3(e), subject #3).

We computed the Dice coefficient on a cohort of subjects only from NMM and a cohort of subjects only from NPH for the methods and report the results in Tables 1 and 2, respectively. We note that VenSeg2 performed worse than VenSeg1 on NPH data set despite having more training data (see Table 2). We used a paired Wilcoxon signed-rank test [21] to compare the methods. For the results on the NMM testing images, we found no significant differences between VenSeg1 and VenSeg3 in terms of Dice coefficients. Both networks performed significantly better ($p < 0.001$) than FreeSurfer and RUDOLPH on the lateral ventricles and the 3rd ventricle, and better than FreeSurfer on the 4th ventricle. For the results on the NPH image testing set, VenSeg3 performed significantly better ($p < 0.001$) than all the other methods on all the ventricle labels.

Table 1. The mean Dice coefficient (and standard deviation) over 25 T1w images from Neuromorphometrics. Ventricular system key: Merged four ventricle labels (Whole), right lateral ventricle (RLV), left lateral ventricle (LLV), third ventricle (3rd), and fourth ventricle (4th). The asterisks mean significantly different (p-value <0.001) to VenSeg1 and VenSeg3.

	Whole	RLV	LLV	3rd	4th
FreeSurfer	0.843*($\pm$0.04)	0.848*($\pm$0.04)	0.848*($\pm$0.04)	0.700*($\pm$0.12)	0.760*($\pm$0.04)
JLF	0.881($\pm$0.03)	0.879($\pm$0.03)	0.888($\pm$0.03)	0.796($\pm$0.04)	0.844($\pm$0.03)
RUDOLPH	0.883*($\pm$0.03)	0.883*($\pm$0.03)	0.888*($\pm$0.03)	0.777*($\pm$0.08)	0.839($\pm$0.04)
VenSeg1	0.902($\pm$0.03)	0.903($\pm$0.03)	0.907($\pm$0.03)	0.821($\pm$0.07)	0.844($\pm$0.04)
VenSeg3	0.902($\pm$0.03)	0.904($\pm$0.03)	0.907($\pm$0.03)	0.817($\pm$0.07)	0.842($\pm$0.04)

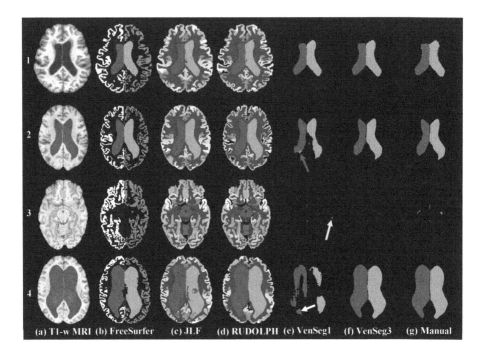

Fig. 3. Segmentation results from three state-of-the-art methods (FreeSurfer, JLF, and RUDOLPH) and two proposed deep networks (VenSeg1 and VenSeg3) compared with a manual rater (column g). Subject #1: T1w image and segmentation results from NMM data set. Subjects #2–4: T1w images and segmentation results from NPH data set, showing moderate, mile, and severe cases. The red arrow in (e2) shows the right lateral ventricle inaccurately labeled as the left lateral ventricle. The yellow arrow in (e3) points to cortex mislabeled as the 4th ventricle. The white arrow in (e4) points to the right ventricle mislabeled as the 3rd ventricle. (Color figure online)

Table 2. The mean Dice coefficient (and standard deviation) over the 70 testing images from the NPH data set. Bold: VenSeg3 is significantly better (p-value <0.001) than the other five methods on all the labels.

	Whole	RLV	LLV	3rd	4th
FreeSurfer	0.937(±0.03)	0.942(±0.03)	0.939(±0.03)	0.840(±0.06)	0.730(±0.08)
JLF	0.930(±0.04)	0.931(±0.05)	0.933(±0.04)	0.865(±0.06)	0.862(±0.04)
RUDOLPH	0.942(±0.05)	0.943(±0.05)	0.944(±0.05)	0.875(±0.07)	0.838(±0.06)
VenSeg1	0.833(±0.14)	0.839(±0.15)	0.832(±0.15)	0.727(±0.21)	0.787(±0.11)
VenSeg2	0.482(±0.24)	0.484(±0.25)	0.480(±0.25)	0.275(±0.28)	0.684(±0.18)
VenSeg3	**0.971**(±0.02)	**0.971**(±0.02)	**0.974**(±0.02)	**0.915**(±0.06)	**0.903**(±0.04)

4 Discussion and Conclusions

We present a 3D U-Net architecture to segment and label the ventricular system in patients with enlarged ventricles. We trained three models on two different data sets using manual delineations as training data. The models were evaluated on 25 NMM subjects and 70 NPH patients and compared to FreeSurfer, JLF, and RUDOLPH.

The model trained on 13 NMM data showed improvement over the state-of-the-art segmentation methods in terms of overlap with expert delineations on the same data set. However, it showed poor performance on the NPH data set, even on images with ventricle size similar to the training data. The segmentation results from this model on subjects #1 and #2 were inconsistent. The model failed to identify the boundary of the lateral ventricles and mislabeled portions of the right ventricle as left on subject #2 (see the red arrow in Fig. 3(e2)). This failure occurred despite the fact that the size of the ventricles in subject #2 is very similar to the ventricle size of subject #1 from NMM. In some cases with small ventricular volume, the model mislabeled the cortex as ventricle (see the yellow arrow in Fig. 3(e3)). In severe cases of NPH, this model cannot handle the pathology as its training data set does not include similar examples; Furthermore it labeled posterior portions of the right ventricle as the 3rd ventricle (see the white arrow in Fig. 3(e4)).

The second network was trained on 38 NMM images, including elderly subjects with enlarged ventricles, since more training data could potentially improve the performance. However, this network provided worse segmentation results than the first one when evaluated on NPH patients. One possible explanation is that adding more training data made the network overfitted on the NMM data set.

The failure of these two networks on NPH patients indicates that the network did not learn only the intensity and spatial information from the training data, since the first network successfully segmented a subject from NMM but failed on a subject with similar ventricle size from the NPH data set. The dominant features learned by the network—that are driving the segmentation—remain a mystery.

The third network was trained on 38 images from both data sets. It performed significantly better than all of the other methods on the entire testing data set, demonstrating both the robustness of the network to high variations of ventricle sizes, but also the importance of careful training data selection for deep learning methods.

References

1. Adams, R., Fisher, C., Hakim, S., Ojemann, R., Sweet, W.: Symptomatic occult hydrocephalus with normal cerebrospinal-fluid pressure: a treatable syndrome. N. Engl. J. Med. **273**(3), 117–126 (1965)
2. de Brebisson, A., Montana, G.: Deep neural networks for anatomical brain segmentation. In: Proceedings of the IEEE Conference on Computer Vision and Pattern Recognition Workshops, pp. 20–28 (2015)

3. Carass, A., et al.: Whole brain parcellation with pathology: validation on ventricu-lomegaly patients. In: Wu, G. (ed.) Patch-MI 2017. LNCS, vol. 10530, pp. 20–28. Springer, Cham (2017). https://doi.org/10.1007/978-3-319-67434-6_3

4. Dice, L.R.: Measures of the amount of ecologic association between species. Ecology **26**(3), 297–302 (1945)

5. Ellingsen, L.M., Roy, S., Carass, A., Blitz, A.M., Pham, D.L., Prince, J.L.: Seg-mentation and labeling of the ventricular system in normal pressure hydrocephalus using patch-based tissue classification and multi-atlas labeling. In: Proceedings of SPIE–the International Society for Optical Engineering, vol. 9784 (2016)

6. Fischl, B.: Freesurfer. NeuroImage **62**(2), 774–781 (2012)

7. Fonov, V.S., Evans, A.C., McKinstry, R.C., Almli, C., Collins, D.: Unbiased nonlin-ear average age-appropriate brain templates from birth to adulthood. NeuroImage **47**, S102 (2009)

8. He, K., Zhang, X., Ren, S., Sun, J.: Deep residual learning for image recognition. arXiv preprint arXiv:1512.03385 (2015)

9. He, K., Zhang, X., Ren, S., Sun, J.: Identity mappings in deep residual networks. arXiv preprint arXiv:1603.05027 (2016)

10. Hebb, A.O., Cusimano, M.D.: Idiopathic normal pressure hydrocephalus: a sys-tematic review of diagnosis and outcome. Neurosurgery **49**(5), 1166–1186 (2001)

11. Ishikawa, M., et al.: Guidelines for management of idiopathic normal pressure hydrocephalus. Neurol. Med.-Chir. **48**(Suppl.), S1–S23 (2008)

12. Kamnitsas, K., et al.: Efficient multi-scale 3D CNN with fully connected CRF for accurate brain lesion segmentation. Med. Image Anal. **36**, 61–78 (2017)

13. Kayalibay, B., Jensen, G., van der Smagt, P.: CNN-based segmentation of medical imaging data. arXiv preprint arXiv:1701.03056 (2017)

14. Kingma, D.P., Ba, J.: Adam: a method for stochastic optimization. arXiv preprint arXiv:1412.6980 (2014)

15. Ledig, C., et al.: Robust whole-brain segmentation: application to traumatic brain injury. Med. Image Anal. **21**(1), 40–58 (2015)

16. Ronneberger, O., Fischer, P., Brox, T.: U-Net: convolutional networks for biomed-ical image segmentation. In: Navab, N., Hornegger, J., Wells, W.M., Frangi, A.F. (eds.) MICCAI 2015. LNCS, vol. 9351, pp. 234–241. Springer, Cham (2015). https://doi.org/10.1007/978-3-319-24574-4_28

17. Roy, S., Butman, J.A., Pham, D.L.: Alzheimers disease neuroimaging initiative, others: robust skull stripping using multiple MR image contrasts insensitive to pathology. NeuroImage **146**, 132–147 (2017)

18. Tustison, N.J., et al.: N4ITK: improved N3 bias correction. IEEE Trans. Med. Imag. **29**(6), 1310–1320 (2010)

19. Ulyanov, D., Vedaldi, A., Lempitsky, V.: Instance normalization: the missing ingre-dient for fast stylization. arXiv preprint arXiv:1607.08022 (2016)

20. Wang, H., Suh, J.W., Das, S.R., Pluta, J.B., Craige, C., Yushkevich, P.A.: Multi-atlas segmentation with joint label fusion. IEEE Trans. Patt. Anal. Mach. Intell. **35**(3), 611–623 (2013)

21. Wilcoxon, F.: Individual comparisons by ranking methods. Biom. Bull. **1**(6), 80–83 (1945)

22. Xu, B., Wang, N., Chen, T., Li, M.: Empirical evaluation of rectified activations in convolutional network. arXiv preprint arXiv:1505.00853 (2015)

Vulnerability Analysis of Chest X-Ray Image Classification Against Adversarial Attacks

Saeid Asgari Taghanaki[1](✉), Arkadeep Das[1,2], and Ghassan Hamarneh[1]

[1] School of Computing Science, Simon Fraser University, Burnaby, Canada
{sasgarit,arkadeep_das,hamarneh}@sfu.ca
[2] Department of Mathematics, Indian Institute of Technology, Guwahati, India

Abstract. Recently, there have been several successful deep learning approaches for automatically classifying chest X-ray images into different disease categories. However, there is not yet a comprehensive vulnerability analysis of these models against the so-called adversarial perturbations/attacks, which makes deep models more trustful in clinical practices. In this paper, we extensively analyzed the performance of two state-of-the-art classification deep networks on chest X-ray images. These two networks were attacked by three different categories (ten methods in total) of adversarial methods (both white- and black-box), namely gradient-based, score-based, and decision-based attacks. Furthermore, we modified the pooling operations in the two classification networks to measure their sensitivities against different attacks, on the specific task of chest X-ray classification.

Keywords: Adversarial perturbation · Chest X-ray classification
Deep learning

1 Introduction

The chest X-ray is among the top most commonly accessible medical imaging examinations used for affordable screening and diagnosis of numerous lung ailments including pneumothorax, mass, cardiomegaly, effusion, and pneumonia. Owing to huge numbers of patients and increasing burden of lung ailments, the workload of radiologists has significantly multiplied. Hence, with an intention to accelerate/support the predictions of radiologists, many machine (deep) learning classification frameworks have emerged over the past few years.

The availability of a new large scale chest X-ray dataset namely "ChestX-ray14" [20], which comprises 30,805 patients and 112,120 chest X-ray images, makes it feasible to apply deep learning without a need for data augmentation or synthetic data. Recently, different standard classification deep networks (AlexNet [7], VGGNet [14] and ResNet [5]) have been applied to this dataset. Wang et al. [20] applied pre-trained AlexNet, GoogLeNet [17], VGG, and ResNet-50 architectures to classify 8 disease categories. They showed that

© Springer Nature Switzerland AG 2018
D. Stoyanov et al. (Eds.): MLCN 2018/DLF 2018/iMIMIC 2018, LNCS 11038, pp. 87–94, 2018.
https://doi.org/10.1007/978-3-030-02628-8_10

ResNet-50 achieves superior performance compared to the other applied models. Guendel et al. [4] proposed a local aware dense network for classification of 14 pathology classes in the ChestX-ray14 dataset. Rajpurkar et al. [12] proposed CheXNet, a 121-layer convolutional neural network trained on ChestX-ray14 for the pneumonia disease detection task, which exceeds average radiologist performance on the F1 metric. Baltruschat et al. [1] proposed a fine-tuned ResNet-50 network which achieved high accuracy on 4 out of the 14 disease classes in the chest X-ray dataset. Yao et al. [21] presented a partial solution to constraints in using LSTMs to leverage inter-dependencies among target labels in predicting 14 pathological classes from chest X-rays.

The *generalizability* of the deep learning methods, i.e. how they perform on unseen chest X-ray test images, have been explored in the above mentioned works to some extent. However, discovery of "adversarial examples" has exposed serious vulnerabilities in even state-of-the-art deep learning systems [8]. As of writing there is no comprehensive study on the *vulnerability* analysis of the state-of-the art classification networks against adversarial perturbations for chest X-rays. Finlayson et al. [3] considered a single attack, namely projected gradient descent [6, 9] on chest X-ray images.

Adversarial images are crafted by adding perturbations, imperceptible to the naked eye, to the clean images to fool machine learning models. Different categories [22] of adversarial attacks on images have been recently developed which have been highly successful in fooling deep neural networks. In the medical image analysis domain, attacks may originate during data-transfer through the Internet or local networks [3]. Even in the case of complete protection from adversarial attacks, training existing deep models with adversarial examples or designing defense mechanisms [23] can improve model generalizability and resilience. In this paper, we present a comprehensive analysis of ten different adversarial attacks on classification of chest X-ray images and investigate how two different standard deep neural networks perform against adversarial perturbations. We perform both white (i.e. producing perturbed images using network A and classifying them by the same network) and black-box (i.e. producing perturbed images using network A and classifying them by network B) attacks.

2   Methods

2.1   Applied Deep Networks

We use two state-of-the-art deep models i.e. Inception-ResNet-v2 [16] and Nasnet-Large [24] to evaluate their performance on classification of both clean and perturbed chest X-ray images. Next, we modify the networks by replacing max-pooling operations with average-pooling to analyze whether the modified networks, especially the ones that are based on single/few pixel perturbation, are less sensitive to attacks. We hypothesize that average-pooling may be more resilient to attacks as it captures more global contextual information from the field of view, instead of selecting a single pixel candidate as max-pooling does.

2.2 Applied Adversarial Attacks

We applied three different categories of attacks namely gradient-based, score-based, and decision-based:

- **Gradient-based** attacks linearize the loss (in our case binary cross-entropy) around an input to which the model predictions for a particular class are most sensitive to. These attacks perturb the image with the gradient of the loss w.r.t. the clean image, gradually and efficiently increasing the magnitude until the model predicts a different label for the perturbed image. In our experiments, we have selected five different gradient-based attacks namely, Fast Gradient Sign Method (G1) [8], Projected Gradient Descent (G2) [9], Deep-Fool (G3) [10], Linfinity Basic Iterative Method (G4) [8],Limited-memory Broyden–Fletcher–Goldfarb–Shanno Method (L-BFGS) (G5) [18] and we demonstrate how the models trained on clean images perform against the crafted adversarial examples.
- **Score-based** attacks rely on confidence scores e.g. softmax class probabilities or logits to numerically estimate the gradient. From this group, we apply Local Search (S1) [11] (a black box attack based on the greedy local search algorithm to find pixels for which the model is the most sensitive and perturbing them to misclassify the input) and the Single Pixel (S2) [15] attacks.
- **Decision-based** attacks [2] solely rely on the predicted class or label of the model without requiring gradients or logits. From this group, we applied Gaussian Blur (D1), Contrast Reduction (D2) and Additive Gaussian Noise (D3) in our experiments. In all of the aforementioned attacks, a line-search is performed internally to find minimal perturbations required by the image to turn it into an adversarial example.

We trained both the networks from scratch with a batch size of 32 and 8 for training the Inception-ResNet-v2 and Nasnet-Large, respectively. RMSProp optimizer [19] with a decay of 0.9 and $\epsilon = 1$ and an initial learning rate of 0.045, decayed every 4 epochs using an exponential rate of 0.94 were used for all of our experiments as described in [16,24]. We set all attack parameters as proposed by their authors and utilized Foolbox [13], to craft adversarial examples.

3 Dataset

We use ChestX-ray14 dataset [20] which comprises 112,120 gray-scale images with 14 disease labels and 1 no-finding label. We treat all the disease classes as positive and formulate a binary classification task of "disease" vs. "non-disease". We randomly selected 95,128 images for training and 16,792 for validation. We randomly picked 200 unseen images as the test set, with 93 images with chest disease labels and 107 having "no finding" labels. These clean images are used for carrying out different adversarial attacks and the models trained on clean images are evaluated against them.

4   Results and Discussion

Figure 1 shows the perturbed images produced by the ten different applied attacks. In Fig. 2, we visualize a few samples where the perturbations are perceptible by human. We observed that most of the produced images by D1 (i.e Gaussian blur), D2 (i.e. contrast reduction), D3 (i.e additive Gaussian noise), S1 (i.e. local search) attack can be easily detected by the naked eye. We also found that S1 requires relatively more time compared to other methods to find an adversarial image.

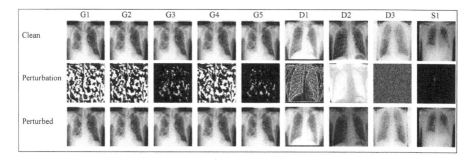

Fig. 1. Perturbed images produced by 10 (3 categories) different attacks.

In Tables 1 and 2, we report accuracy and area under ROC for two networks with/without modification for clean and ten different adversarial attacks (white- and black-box). Note that the single pixel attack [15] i.e. S2 (from the score based attacks category) failed to fool the networks for the entire test set which shows the single pixel attack works well on RGB (colored images) but not on gray-scale X-ray images as it is not simple to fool a deep model by changing only a single "gray-scale" pixel.

Table 1. Performance of original/modified Inception-ResNet-v2 (IR2) and Nasnet-Large (NL) against ten different *white-box* attacks. In the Table, MP, AP, Acc., AU refer to max-pooling, average-pooling, accuracy, and area under ROC, respectively.

	Model	Metrics	Clean	Gradient					Decision			Score	
				G1	G2	G3	G4	G5	D1	D2	D3	S1	S2
MP	IR2	Acc.	0.70	0.00	0.00	0.00	0.00	0.00	0.04	0.10	0.32	0.65	0.70
		AU	0.75	0.00	0.00	0.00	0.00	0.00	0.06	0.19	0.52	0.74	0.75
	NL	Acc.	0.73	0.00	0.00	0.01	0.00	0.00	0.06	0.41	0.30	0.32	0.73
		AU	0.77	0.00	0.00	0.10	0.00	0.00	0.10	0.66	0.58	0.55	0.77
AP	IR2	Acc.	0.71	0.00	0.00	0.00	0.00	0.00	0.04	0.24	0.14	0.62	0.71
		AU	0.74	0.00	0.00	0.00	0.00	0.00	0.06	0.39	0.26	0.72	0.74
	NL	Acc.	0.72	0.00	0.00	0.00	0.00	0.00	0.03	0.41	0.48	0.72	0.72
		AU	0.74	0.00	0.00	0.00	0.00	0.00	0.04	0.64	0.64	0.74	0.74

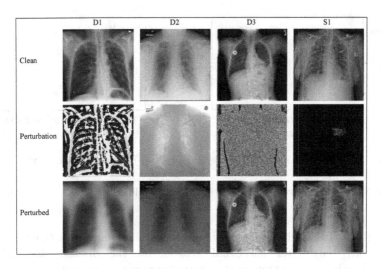

Fig. 2. Human perceptible adversarial perturbations

Table 2. Performance of original/modified Inception-ResNet-v2 (IR2) and Nasnet-Large (NL) against ten different *black-box* attacks. In the Table, MP, AP, Acc., AU refer to max-pooling, average-pooling, accuracy, and area under ROC, respectively.

	Model	Metrics	Clean	Gradient					Decision			Score	
				G1	G2	G3	G4	G5	D1	D2	D3	S1	S2
MP	IR2	Acc.	0.70	0.46	0.43	0.43	0.43	0.43	0.53	0.81	0.36	0.45	0.70
		AU	0.75	0.44	0.41	0.40	0.41	0.41	0.43	0.84	0.24	0.40	0.75
	NL	Acc.	0.73	0.53	0.51	0.51	0.51	0.51	0.57	0.58	0.74	0.56	0.73
		AU	0.77	0.52	0.49	0.49	0.49	0.49	0.51	0.55	0.82	0.55	0.77
AP	IR2	Acc.	0.71	0.51	0.52	0.52	0.52	0.51	0.53	0.29	0.40	0.53	0.71
		AU	0.74	0.49	0.49	0.49	0.47	0.50	0.47	0.24	0.40	0.52	0.74
	NL	Acc.	0.72	0.59	0.58	0.58	0.58	0.58	0.49	0.53	0.51	0.38	0.72
		AU	0.74	0.59	0.58	0.58	0.58	0.58	0.46	0.52	0.46	0.39	0.74

As reported in Table 1, the gradient based attacks were almost completely successful in fooling both networks (with/without modification) when the victim model for attack was the same reference model, i.e. in a white-box attack scenario. The decision and score based attacks were almost unsuccessful in fooling the models. We observed that Nasnet-Large with average pooling was 18% stronger in comparison to Nasnet-Large with max pooling. Note that the local search attack (S1) completely failed against Nasnet-Large with average-pooling.

In Table 2, we show the performance of both the networks against the black-box attacks i.e. we craft adversarial images with Inception-ResNet-v2, but test them with Nasnet-Large and vice versa. As reported in the table, almost all the methods were partially successful but not as high as white-box attacks. For gradient based black-box attacks, average pooling shows more resiliency against the attacks. We observed that for 23% ± 9% and 27% ± 8% of the test samples both the networks failed on the same cases for average and max pooling, respectively.

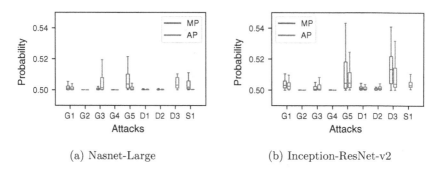

(a) Nasnet-Large   (b) Inception-ResNet-v2

Fig. 3. Probability values of the original and modified networks after attack

In Fig. 3, we show the probability values of the two networks (with/without modification) only after successful attacks on the disease class. Higher ranges/values indicate a stronger attack or a more vulnerable network. As shown in the figure, D3 (i.e. Additive Gaussian Noise), S1 (i.e. Local search) and G5 (i.e. L-BFGS) attacks are highly sensitive to the choice of pooling (max/average) operation. The range of the attack's confidence varies from $\sim$0.50 to $\sim$0.55. Note that absence of a box in the figure means there was no successful attack for a disease class in that experiment. In Fig. 4, we visualize the accuracy of Inception-ResNet-v2 for different groups of attacks and, in the same plot, we show the perceptibility of each group of the attacks (i.e. the difficulty level for a human to detect a perturbed image). Note that lower accuracy and harder detection (lower right of the plot) implies a more successful attack. As shown in the figure, gradient based attacks are the most successful ones in terms of fooling both human (i.e. perception) and machine (i.e. accuracy).

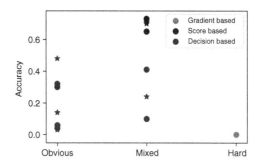

Fig. 4. Distribution of adversarial images crafted based on Inception-ResNet-v2 network w.r.t. human perception. The dots and stars in the figure refer to max and average pooling, respectively. The words Obvious, Mixed, and Hard refer to the level of difficulty for a human to perceive the attacks.

5   Conclusion

In this paper, we extensively tested the vulnerability of the two state-of-the-art deep classification networks against ten different adversarial attacks on chest X-ray images. We found that the single pixel attack completely failed for gray-level X-ray chest images. We also showed that the pooling operation can make a considerable difference for some attacks, even leading to a complete failure of the attack for a particular class. We also demonstrated that the crafted adversarial images with some of the attacks, e.g. Gaussian blur and contrast reduction methods, can be simply detected with the naked eye. Finally, we showed that the gradient based attacks applied to the chest X-ray images are the most successful in terms of fulling both machine and human. Although both networks, Inception-ResNet-v2 and Nasnet-Large, failed against gradient-based attacks, in general, the latter (with average pooling) was more resilient to decision and score based attacks.

Acknowledgments. We thank NVIDIA Corporation for GPU donation and MITACS Globalink for funding.

References

1. Baltruschat, I.M., Nickisch, H., Grass, M., Knopp, T., Saalbach, A.: Comparison of deep learning approaches for multi-label chest X-ray classification. arXiv preprint arXiv:1803.02315 (2018)
2. Brendel, W., Rauber, J., Bethge, M.: Decision-based adversarial attacks: reliable attacks against black-box machine learning models. arXiv preprint arXiv:1712.04248 (2017)
3. Finlayson, S.G., Kohane, I.S., Beam, A.L.: Adversarial attacks against medical deep learning systems. arXiv preprint arXiv:1804.05296 (2018)
4. Guendel, S., et al.: Learning to recognize abnormalities in chest X-rays with location-aware dense networks. arXiv preprint arXiv:1803.04565 (2018)
5. He, K., Zhang, X., Ren, S., Sun, J.: Deep residual learning for image recognition. In: Proceedings of the IEEE Conference on Computer Vision and Pattern Recognition, pp. 770–778 (2016)
6. Kannan, H., Kurakin, A., Goodfellow, I.: Adversarial logit pairing. arXiv preprint arXiv:1803.06373 (2018)
7. Krizhevsky, A., Sutskever, I., Hinton, G.E.: Imagenet classification with deep convolutional neural networks. In: Advances in Neural Information Processing Systems, pp. 1097–1105 (2012)
8. Kurakin, A., Goodfellow, I., Bengio, S.: Adversarial examples in the physical world. arXiv preprint arXiv:1607.02533 (2016)
9. Madry, A., Makelov, A., Schmidt, L., Tsipras, D., Vladu, A.: Towards deep learning models resistant to adversarial attacks. arXiv preprint arXiv:1706.06083 (2017)
10. Moosavi-Dezfooli, S.M., Fawzi, A., Frossard, P.: DeepFool: a simple and accurate method to fool deep neural networks. In: Proceedings of the IEEE Conference on Computer Vision and Pattern Recognition, pp. 2574–2582 (2016)
11. Narodytska, N., Kasiviswanathan, S.P.: Simple black-box adversarial perturbations for deep networks. arXiv preprint arXiv:1612.06299 (2016)

12. Rajpurkar, P., et al.: Chexnet: radiologist-level pneumonia detection on chest X-rays with deep learning. arXiv preprint arXiv:1711.05225 (2017)
13. Rauber, J., Brendel, W., Bethge, M.: Foolbox v0. 8.0: a python toolbox to benchmark the robustness of machine learning models. arXiv preprint arXiv:1707.04131 (2017)
14. Simonyan, K., Zisserman, A.: Very deep convolutional networks for large-scale image recognition. arXiv preprint arXiv:1409.1556 (2014)
15. Su, J., Vargas, D.V., Kouichi, S.: One pixel attack for fooling deep neural networks. arXiv preprint arXiv:1710.08864 (2017)
16. Szegedy, C., Ioffe, S., Vanhoucke, V., Alemi, A.A.: Inception-v4, inception-ResNet and the impact of residual connections on learning. In: AAAI, vol. 4, p. 12 (2017)
17. Szegedy, C., et al.: Going deeper with convolutions. In: Proceedings of the IEEE Conference on Computer Vision and Pattern Recognition, pp. 1–9 (2015)
18. Szegedy, C., et al.: Intriguing properties of neural networks. arXiv preprint arXiv:1312.6199 (2013)
19. Toshev, A., Szegedy, C.: DeepPose: human pose estimation via deep neural networks. In: Proceedings of the IEEE Conference on Computer Vision and Pattern Recognition, pp. 1653–1660 (2014)
20. Wang, X., Peng, Y., Lu, L., Lu, Z., Bagheri, M., Summers, R.M.: ChestX-ray8: hospital-scale chest X-ray database and benchmarks on weakly-supervised classification and localization of common thorax diseases. In: 2017 IEEE Conference on Computer Vision and Pattern Recognition (CVPR), pp. 3462–3471. IEEE (2017)
21. Yao, L., Poblenz, E., Dagunts, D., Covington, B., Bernard, D., Lyman, K.: Learning to diagnose from scratch by exploiting dependencies among labels. arXiv preprint arXiv:1710.10501 (2017)
22. Yuan, X., He, P., Zhu, Q., Bhat, R.R., Li, X.: Adversarial examples: attacks and defenses for deep learning. arXiv preprint arXiv:1712.07107 (2017)
23. Zantedeschi, V., Nicolae, M.I., Rawat, A.: Efficient defenses against adversarial attacks. In: Proceedings of the 10th ACM Workshop on Artificial Intelligence and Security, pp. 39–49. ACM (2017)
24. Zoph, B., Vasudevan, V., Shlens, J., Le, Q.V.: Learning transferable architectures for scalable image recognition. arXiv preprint arXiv:1707.07012, vol. 2, no. 6 (2017)

First International Workshop on Interpretability of Machine Intelligence in Medical Image Computing, iMIMIC 2018

Collaborative Human-AI (CHAI): Evidence-Based Interpretable Melanoma Classification in Dermoscopic Images

Noel C. F. Codella[1]([⊠]), Chung-Ching Lin[1], Allan Halpern[2], Michael Hind[1], Rogerio Feris[1], and John R. Smith[1]

[1] IBM T.J. Watson Research Center, Yorktown Heights, NY 10598, USA
{nccodell,cclin,hindm,rferis,jsmith}@us.ibm.com
[2] Memorial Sloan-Kettering Cancer Center, New York, NY 10065, USA
halperna@mskcc.org

Abstract. Automated dermoscopic image analysis has witnessed rapid growth in diagnostic performance. Yet adoption faces resistance, in part, because no evidence is provided to support decisions. In this work, an approach for evidence-based classification is presented. A feature embedding is learned with CNNs, triplet-loss, and global average pooling, and used to classify via kNN search. Evidence is provided as both the discovered neighbors, as well as localized image regions most relevant to measuring distance between query and neighbors. To ensure that results are relevant in terms of both label accuracy and human visual similarity for any skill level, a novel hierarchical triplet logic is implemented to jointly learn an embedding according to disease labels and non-expert similarity. Results are improved over baselines trained on disease labels alone, as well as standard multiclass loss. Quantitative relevance of results, according to non-expert similarity, as well as localized image regions, are also significantly improved.

Keywords: Deep learning · Evidence · Explainable · Interpretable
Triplet-loss · Global average pooling · Weighted activation maps
Dermoscopy · Melanoma

1   Introduction

In the past decade, advancement in computer vision techniques has been facilitated by both large-scale datasets and deep learning approaches. Now this trend is influencing dermoscopic image analysis, where the International Skin Imaging Collaboration (ISIC) has organized a large public repository of high quality annotated images, referred to as the ISIC Archive (http://isic-archive.com). From this repository, snapshots of the dataset have been used to host two consecutive years of benchmark challenges [1,2], which have increased interest in the computer vision community [2–6], and supported the development of methods that surpassed the diagnostic performance of expert clinicians [2–4]. However,

© Springer Nature Switzerland AG 2018
D. Stoyanov et al. (Eds.): MLCN 2018/DLF 2018/iMIMIC 2018, LNCS 11038, pp. 97–105, 2018.
https://doi.org/10.1007/978-3-030-02628-8_11

despite these advancements, deployment to clinical practice remains problematic, in part, because most systems lack evidence for predictions that can be interpreted by users of varying skill.

Recent works have attempted to provide various forms of evidence for decisions. Methods to visualize feature maps in neural networks were introduced in 2015 [7], facilitating better understanding of the behavior of networks, but not justifying predictions made on specific image inputs. Global average pooling approaches have been proposed [8], which get closer to justifying decisions on specific image inputs by indicating importance of image regions to those decisions, but fail to provide specific evidence behind the classifications.

An extensive body of prior work around content-based image retrieval (CBIR) is perhaps the most relevant toward providing classification decisions with evidence [9–13]. Early approaches relied on low-level features and bag-of-visual words, [9–11], but suffered from the "semantic gap": feature similarity did not necessarily correlate to label similarity. Later approaches have used deep neural networks to learn an embedding for search, reducing semantic gap issues [13]. However, such methods have still suffered from a "user-gap": what an embedding learns to consider as similar from disease point-of-view does not necessarily correlate with human measures of similarity. In addition, users cannot determine what spatial regions of images contributed most to distance measures.

Specific to the domain of dermoscopic image analysis, one work proposed to learn and localize clinically discriminative patterns in images [5]; however, this output can only be verified by experts who know how to identify the patterns. In addition, classifier decision localization has been proposed for multimodal systems [14]; however, localization information alone isn't sufficient as evidence for classification decisions.

In this work, a solution for a Collaborative Human-AI (CHAI) dermoscopic image analysis system is presented. In order to facilitate interpretability of evidence by clinical staff of any skill level, this approach (1) introduces a novel hierarchical triplet loss to learn an embedding for k-nearest neighbor search, optimized jointly from disease labels as well as non-expert human similarity, and (2) provides localization information in the form of *query-result activation map pairs*, which designate regions in query and result images used to measure distance between the two. Experiments demonstrate that the proposed approach improves classification performance in comparison to models trained on disease labels alone, as well as models trained with classification loss. The relevancy of results, according to non-expert similarity, are also significantly improved.

2   Methods

2.1   Triplet-Loss with Global Average Pooling

The proposed embedding framework is displayed in Fig. 1a. A triplet loss structure [15] is combined with penultimate global average pooling layers [8] to learn a discriminative feature embedding that supports activation localization. AlexNet, including up to the "conv5" layer, is used as the CNN.

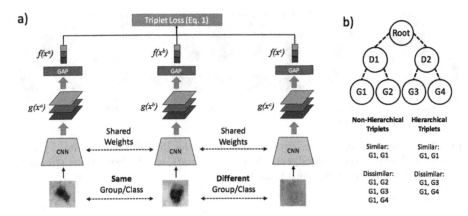

Fig. 1. (a) Proposed triplet loss framework with global average pooling (GAP) architecture. (b) Top: Visual example of proposed hierarchical annotation groups. The first level grouping is by disease label (D1-2), and the second level by human visual similarity (G1-4). Bottom: Example triplet logic is shown as pairing between groups.

In order to train, 3 deep neural networks with shared weights across 3 input images (x^a, x^b, x^c) produce feature embeddings $(f(x^a), f(x^b), f(x^c))$. The following objective function over those embeddings provides the gradient for backpropagation:

$$L = max \left[0, l + D(f(x^a), f(x^b)) - \frac{1}{2}(D(f(x^a), f(x^c)) + D(f(x^b), f(x^c))) \right] \quad (1)$$

where $D()$ is a distance metric (squared Euclidean distance), l is a constant representing the margin (set to 1), x^a and x^b are considered similar inputs, and x^c is a dissimilar input.

The feature embedding is comprised of a global average pooling (GAP) layer to support generation of a *query-result activation map pair*, which highlights regions of pairs of images that contributed most toward the distance measure between them. This is done by combining the feature layer activation maps prior to global average pooling into a single grayscale image, weighted by the squared differences between two image feature embeddings:

$$A^q(i,j) = \sum_{z=0}^{d} g_z(x^q, i, j)) \cdot (f_z(x^q) - f_z(x^r))^2 \quad (2)$$

where $A^q(i,j)$ is the query activation map (QAM), $g_z(x,i,j)$ is the zth filter bank before global average pooling, d is the dimensionality of the filter bank, x^q is the query image, x^r is a search result image, and:

$$f_z(x) = \frac{1}{n^2} \sum_{i=0}^{n} \sum_{j=0}^{n} g_z(x, i, j) \quad (3)$$

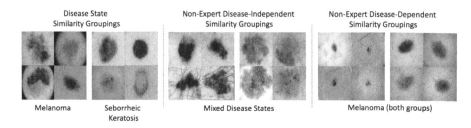

Fig. 2. Example groups by disease category from the ISIC database (left), by non-expert similarity disregarding disease diagnosis (center), and by non-expert similarity constrained within disease groups (right).

is the zth feature embedding element. The result activation map (RAM) A^r in the query-result pair is likewise computed as in Eq. 2, where $g_z(x^q, i, j)$ is replaced with $g_z(x^r, i, j)$.

2.2 Hierarchical Triplet Selection Logic

An example of the hierarchical triplet selection logic is shown in Fig. 1b. Given visually similar groups annotated under disease labels, a hierarchical selection process pairs images as similar if they are siblings within the same group under a disease parent. Dissimilar images include images from other disease states, but exclude cousin images (images within the same disease, but different similarity group). A non-hierarchical selection process takes dissimilar images from any other group, including cousins.

2.3 Experimental Design

The 2017 International Skin Imaging Collaboration (ISIC) challenge on Skin Lesion Analysis Toward Melanoma Detection [1] dataset is used for experimentation. This is a public dataset consisting of 2000 training dermoscopic images and 600 test images. Experiments on this data compare between the following 6 feature embeddings for kNN classification:

Baseline: The first is the 4096 dimensional fc6 feature embedding layer of the AlexNet architecture trained on the CASIA-WebFace dataset, described in prior work [15]. This is used as the baseline as it is one of the only human-skin focused pre-trained networks currently available.

BaselineFT: Baseline 4096 is fine-tuned for disease labels using standard multi-class accuracy loss. This method represents one of the most common approaches for generating embeddings for KNN classification in practice.

Disease: This is a 1024 dimensional CHAI feature embedding, learned from disease labels on the training data partition of the ISIC dataset, fine-tuned from the baseline.

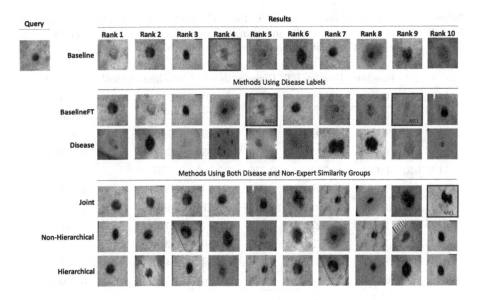

Fig. 3. Example search results across systems, displayed according to similarity rank, with rank 1 being the most similar image in the training dataset. Red borders signify instances of melanoma. (Color figure online)

Joint: This is a CHAI feature embedding jointly fine-tuned from baseline using disease labels, as well as non-expert human similarity groupings, consisting of 1700 images pulled from the ISIC Archive (excluding test images), annotated into 37 distinct groups. The annotator was not given disease labels, and thus may mix diseases within groups. Example groups are shown in Fig. 2.

Hierarchical: This is a CHAI feature embedding fine-tuned from the disease model using human similarity groups that are dependent on disease labels. All 2000 images and 600 test images were annotated from the 2017 ISIC challenge dataset, partitioned into 20 groups of similar images under melanoma, 12 groups under seborrheic keratosis, and 15 groups under benign nevus, according to a non-expert human user. Because this type of data is difficult to annotate, only 1000 training images were used for fine-tuning. The remainder of the data was used for evaluation. Examples of these groups are shown in Fig. 2. Triplets were selected based on hierarchical logic.

Non-hierarchical: To isolate the effects of hierarchical logic, and disease labels being provided to the annotator, the hierarchical groups are used to create triplets using non-hierarchical logic: dissimilar images are selected from any other group, including cousins.

Most learning parameters are kept consistent with prior art [15], including the activation map feature dimensionality of 1024 [8], batch size 128, momentum of 0.9, "step" learning rate policy, learning rate for transferred weights (0.00001), and learning rate for randomly initialized GAP layer (0.01). For BaselineFT, a

Table 1. Melanoma Classification AUC for each method and number of neighbors (k), followed by number of results matching human similarity relevancy (REL), and Jaccard (JA) of QAM against segmentation ground truth.

	Baseline	BaselineFT	Disease	Joint	Non-hierarchical	Hierarchical
AUC k3	0.663	0.700	**0.734**	0.704	0.713	0.729
AUC k5	0.675	0.714	0.744	0.738	0.743	**0.756**
AUC k10	0.681	0.709	0.757	0.754	0.749	**0.774**
AUC k20	0.712	0.745	0.775	0.752	0.769	**0.783**
AUC k40	0.691	0.742	0.776	0.760	0.776	**0.786**
REL k3	0.942	1.005	0.865	1.048	**1.212**	1.125
REL k5	1.505	1.608	1.412	1.678	**1.958**	1.872
REL k10	2.875	3.027	2.632	3.147	**3.793**	3.658
REL k20	5.470	5.772	4.903	6.067	**7.300**	6.968
REL k40	10.283	10.703	9.125	11.507	**13.958**	13.333
JA	NA	NA	0.176	0.201	0.193	**0.208**

learning rate of 0.01 was used for fc8, 0.001 for fc7 and fc6, and 0.00001 for earlier layers. For all triplet experiments, 150,000 triplets were randomly generated for training, and 50,000 triplets for validation.

The area under receiver operating characteristic (ROC) curve (AUC) is used to measure melanoma classification performance on the dataset, according to average vote among returned nearest neighbors. The hierarchical similarity annotations were used to measure the average number of results matching non-expert human relevancy (REL) across all experiments. Finally, the quality of query activation maps are quantitatively measured by comparing the maps against ground truth segmentation according to Jaccard (JA).

3   Results

Table 1 shows the measured AUC for each model type and variable number of neighbors (k), the number of results matching non-expert human similarity relevance (REL), and the Jaccard of the query activation maps as judged against ground truth segmentations. For comparison, standard classification output from multi-class loss used to train *BaselineFT* produces an AUC of 0.772. The top AUC measured for the challenge was 0.874 [5].

For $k = 3$, *Disease* achieved the highest AUC. Surprisingly, at $k = 20, 40$, *Disease* outperforms the classification output of *BaselineFT* (0.772 AUC). For all other values of k, the *Hierarchical* triplet loss embedding achieved the highest performance. At $k = 40$, these performance numbers were comparable with predictive systems submitted to the challenge (rank 11 out of 23 submissions). The *Hierarchical* triplet loss also achieved the second highest number of human similarity relevant results. While the *Non-Hierarchical* method achieved the highest

Disease		Joint		Non-Hierarchical		Hierarchical	
Rank 1	Rank 2	Rank 1	Rank 2	Rank 1	Rank 2	Rank 1	Rank 2

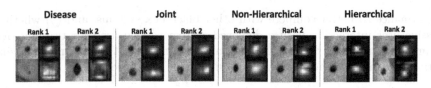

Fig. 4. Example query-result activation pairs for search results. In each group of 4 images: *Top-Left:* query image. *Top-Right:* query activation map. *Bottom-Left:* search result. *Bottom-Right:* search result activation map.

degree of human similarity relevant results, this came at the marginal cost of some classification performance in comparison to *Hierarchical* triplets. However, *Non-Hierarchical* has still matched the classification performance of *Disease*, and outperformed the standard multiclass loss of *BaselineFT*. *Joint* also showed improvements to relevance of human similarity in comparison to *Disease*, but suffered a more harsh penalty to classification performance in comparison to *Hierarchical* and *Non-Hierarchical*.

Representative search results can be inspected in Fig. 3. One can observe here how *Disease*, trained directly on triplets from disease labels, does not translate into the most "relevant" results by human measure: clearly, rank 3 has returned a hypo-pigmented lesion for a pigmented lesion query. In contrast, *Joint*, while maintaining a robust improvement in AUC measures over *Baseline* and *BaselineFT*, has additionally learned to balance disease similarity with a more human measure of similarity. *Hierarchical* has both managed to improve classification performance and human similarity.

Finally, example query-result activation map pairs are shown in Fig. 4. Interestingly, *Disease* learned to examine a broad image extent during comparisons (even potentially irrelevant areas of images), whereas for the models trained with human measures of similarity, the systems have learned to focus more to the localized lesion area. This is confirmed in the over 10% quantitative improvement in Jaccard index comparing to ground truth lesion segmentations, as shown in Table 1.

4 Conclusion

In conclusion, "CHAI", a Collaborative Human-AI system to perform comprehensive evidence-based melanoma classification in dermoscopic images has been presented. Evidence is provided as both the nearest neighbors used for classification, as well as query-result activation map pairs that visualize regions of the images contributing most toward a distance computation. Using a novel hierarchical triplet loss, non-expert human similarity is used to tailor the feature embedding to more closely approximate human judgments of relevance, while simultaneously improving classification performance and the quality of the activation maps. Future work must be carried-out to determine (1) whether the method has the potential to improve adoption, (2) how to improve classification performance to better compete with other black-box systems, and (3) whether passive user interaction with a deployed system can be used for training (for example, from a user clicking on a specific evidence result) to improve classification performance and relevance over time with continued use.

References

1. Codella, N., et al.: Skin lesion analysis toward melanoma detection: a challenge at the international symposium on biomedical imaging (ISBI) 2017, hosted by the international skin imaging collaboration (ISIC). In: IEEE International Symposium of Biomedical Imaging (ISBI) (2018)
2. Marchetti, M., et al.: Results of the 2016 international skin imaging collaboration international symposium on biomedical imaging challenge: comparison of the accuracy of computer algorithms to dermatologists for the diagnosis of melanoma from dermoscopic images. J. Am. Acad. Dermatol. **78**(2), 270–277 (2018)
3. Codella, N.C.F., et al.: Deep learning ensembles for melanoma recognition in dermoscopy images. IBM J. Res. Dev. **61**(4/5), 5:1–5:15 (2017)
4. Esteva, A., et al.: Dermatologist-level classification of skin cancer with deep neural networks. Nature **542**, 115–118 (2017)
5. Menegola, A., Tavares, J., Fornaciali, M., Li, L.T., Avila, S., Valle, E.: RECOD titans at ISIC challenge 2017. In: 2017 International Symposium on Biomedical Imaging (ISBI) Challenge on Skin Lesion Analysis Towards Melanoma Detection. https://arxiv.org/pdf/1703.04819.pdf
6. Diaz, I.G.: Incorporating the knowledge of dermatologists to convolutional neural networks for the diagnosis of skin lesions. In: 2017 International Symposium on Biomedical Imaging (ISBI) Challenge on Skin Lesion Analysis Towards Melanoma Detection. https://arxiv.org/abs/1703.01976
7. Yosinki, J., Clune, J., Nguyen, A., Fuchs, T., Lipson, H.: Understanding neural networks through deep visualization. In: Deep Learning Workshop of International Conference on Machine Learning (ICML) (2015)
8. Zhou, B., Khosla, A., Lapedriza, A., Oliva, A., Torralba, A.: Learning deep features for discriminative localization. In: Computer Vision and Pattern Recognition (CVPR) (2016)
9. Akgul, C.B., Rubin, D.L., Napel, S., Beaulieu, C.F., Greenspan, H., Acar, B.: Content-based image retrieval in radiology: current status and future directions. J. Digit. Imaging **24**(2), 208–222 (2011)
10. Müller, H., Kalpathy–Cramer, J., Caputo, B., Syeda-Mahmood, T., Wang, F.: Overview of the first workshop on medical content–based retrieval for clinical decision support at MICCAI 2009. In: Caputo, B., Müller, H., Syeda-Mahmood, T., Duncan, J.S., Wang, F., Kalpathy-Cramer, J. (eds.) MCBR-CDS 2009. LNCS, vol. 5853, pp. 1–17. Springer, Heidelberg (2010). https://doi.org/10.1007/978-3-642-11769-5_1
11. Ballerini, L., Li, X., Fisher, R.B., Rees, J.: A query-by-example content-based image retrieval system of non-melanoma skin lesions. In: Caputo, B., Müller, H., Syeda-Mahmood, T., Duncan, J.S., Wang, F., Kalpathy-Cramer, J. (eds.) MCBR-CDS 2009. LNCS, vol. 5853, pp. 31–38. Springer, Heidelberg (2010). https://doi.org/10.1007/978-3-642-11769-5_3
12. Li, Z., Zhang, X., Muller, H., Zhang, S.: Large-scale retrieval for medical image analytics: a comprehensive review. Med. Image Anal. **43**, 66–84 (2018)

13. Chung, Y.A., Weng, W.H.: Learning deep representations of medical images using siamese CNNs with application to content-based image retrieval. In: NIPS 2017 Workshop on Machine Learning for Health (ML4H) (2017)
14. Ge, Z., Demyanov, S., Chakravorty, R., Bowling, A., Garnavi, R.: Skin disease recognition using deep saliency features and multimodal learning of dermoscopy and clinical images. In: Descoteaux, M., Maier-Hein, L., Franz, A., Jannin, P., Collins, D.L., Duchesne, S. (eds.) MICCAI 2017. LNCS, vol. 10435, pp. 250–258. Springer, Cham (2017). https://doi.org/10.1007/978-3-319-66179-7_29
15. Zhang, S., et al.: Tracking persons-of-interest via adaptive discriminative features. In: Leibe, B., Matas, J., Sebe, N., Welling, M. (eds.) ECCV 2016. LNCS, vol. 9909, pp. 415–433. Springer, Cham (2016). https://doi.org/10.1007/978-3-319-46454-1_26

Automatic Brain Tumor Grading from MRI Data Using Convolutional Neural Networks and Quality Assessment

Sérgio Pereira[1,2]([✉]), Raphael Meier[3], Victor Alves[2], Mauricio Reyes[4], and Carlos A. Silva[1]([✉])

[1] CMEMS-UMinho Research Unit, University of Minho, Guimarães, Portugal
id5692@alunos.uminho.pt, csilva@dei.uminho.pt
[2] Centro Algoritmi, University of Minho, Braga, Portugal
[3] Support Center for Advanced Neuroimaging,
Institute for Diagnostic and Interventional Neuroradiology,
University Hospital Inselspital and University of Bern, Bern, Switzerland
[4] Institute for Surgical Technology and Biomechanics,
University of Bern, Bern, Switzerland

Abstract. Glioblastoma Multiforme is a high grade, very aggressive, brain tumor, with patients having a poor prognosis. Lower grade gliomas are less aggressive, but they can evolve into higher grade tumors over time. Patient management and treatment can vary considerably with tumor grade, ranging from tumor resection followed by a combined radio- and chemotherapy to a "wait and see" approach. Hence, tumor grading is important for adequate treatment planning and monitoring. The gold standard for tumor grading relies on histopathological diagnosis of biopsy specimens. However, this procedure is invasive, time consuming, and prone to sampling error. Given these disadvantages, automatic tumor grading from widely used MRI protocols would be clinically important, as a way to expedite treatment planning and assessment of tumor evolution. In this paper, we propose to use Convolutional Neural Networks for predicting tumor grade directly from imaging data. In this way, we overcome the need for expert annotations of regions of interest. We evaluate two prediction approaches: from the whole brain, and from an automatically defined tumor region. Finally, we employ interpretability methodologies as a quality assurance stage to check if the method is using image regions indicative of tumor grade for classification.

1 Introduction

Gliomas are the most common primary brain tumors, being graded according to their malignancy. The most aggressive one is Glioblastoma Multiforme (GBM). These high grade gliomas (HGG) proliferate and infiltrate the surrounding tissues at a very fast pace. In fact, patients have a very short life expectancy, even if under treatment [16]. Lower grade gliomas (LGG) are less aggressive, and patients have a better prognosis. Nevertheless, LGG can evolve into HGG, hence,

© Springer Nature Switzerland AG 2018
D. Stoyanov et al. (Eds.): MLCN 2018/DLF 2018/iMIMIC 2018, LNCS 11038, pp. 106–114, 2018.
https://doi.org/10.1007/978-3-030-02628-8_12

follow-up is required [4]. Glioma grading is crucial when deciding the treatment procedure, which can range from surgery followed by chemo- and radiotherapy, to a "wait and see" approach. The latter avoids invasive procedures and is more common with LGG [4,8].

Histopathological diagnosis of biopsy specimens is the gold standard for glioma grading. However, it is time consuming, invasive, and prone to sampling error [17]. MRI is the standard imaging technique for brain tumor diagnosis in clinical practice. In general, attributes of HGG in MRI include the contrast enhancing tumor tissue, necrotic core, edema, non-enhancing tumor, and mass effect. LGG are usually more diffuse, non-enhancing, smaller, and cause less mass effect. Nonetheless, some HGG may have some attributes of LGG, and vice-versa [4,13,16]. Tumor grading from imaging data would be useful in clinical practice, since it would avoid the sampling error, and expedite treatment planning by anticipating the histopathological results [17]. Additionally, it would avoid the invasive biopsy procedures during follow-up. Studies suggest that perfusion MRI is more informative for glioma grading than structural MRI sequences [17]. Still, perfusion MRI is not widely acquired in clinical practice [3]; in fact, perfusion MRI is seen as a plus, while structural MRI is part of the current consensus recommendations for standardized brain tumor imaging [2]. Computer-based tumor grading from MRI is relatively unexplored. Zacharaki et al. [17] predict the grade of gliomas from MRI images using a Support Vector Machine classifier. The method requires radiologists to manually define four regions of interest (ROI) in the tumor. Khawaldeh et al. [6] use convolutional neural networks (CNN) in a semi-automated approach where the tumor grade is predicted from 2D slices selected by radiologists, which may result in multiple and possibly ambiguous predictions for the same patient.

CNNs offer the potential for learning tumor grading directly from imaging data without human-defined ROIs. However, these methods may fall into overfitting, and learn spurious patterns in the data. Hence, a quality assurance stage before deployment of these methods is desirable. As shown by Pereira et al. [9], interpretability of machine learning methods, through explanations of their predictions, allows one to assess which parts of the MRI image are more important for a prediction. In this way, one can evaluate if a model is trustworthy. Moreover, explanations may provide hints on undesirable behaviors, and allow one to devise improving strategies. The contributions in this paper are the following. (i) We propose to use 3D CNN for automatic glioma grading from conventional multisequence MRI, either from the whole brain, or an automatically defined tumor ROI. (ii) We assess the predictions by means of visual explanations. In this way, we were able to assess the predictions' trustworthiness and, as shown in the experiments, detect a problem in pre-processing. Finally, (iii) we validate our approach on a publicly available database, making it more easily comparable with future proposals.

2   Methods

The proposed grading system has two main stages: ROI extraction, and glioma grade prediction. Additionally, we have an interpretation of predictions stage that serves as prediction quality assessment, and we use it for two purposes. First, to evaluate if regions indicative of tumor grade are the most relevant ones for classification. Second, to identify possible problems with the method (e.g. focus on spurious patterns) and devise strategies to obtain better classifiers.

2.1   Extraction of the Region of Interest

We consider and evaluate glioma grading from two ROI: the whole brain, and the tumor region. First, we automatically identify these regions in the image, and define a bounding box around them. Second, these volumes are extracted, resized to a fixed size, and fed into the tumor grade classification CNN. We note that an independent CNN is trained for each of the ROI. Regarding the whole brain region, in a skull-stripped image a bounding box can be easily defined from the brain mask.

For the tumor ROI, a bounding box is defined after segmenting the whole tumor. In order to account for segmentation mistakes, we give a margin of 10 voxels in each side of the bounding box, while maintaining the aspect ratio of the tumor. Segmentation of the whole tumor from multisequence MRI is achieved with a 3D U-net-inspired [10] fully convolutional network; the network architecture is depicted in Fig. 1 (top). A 3D patch is extracted from each MRI sequence, stacked as channels, and fed into the network. The encoder path is responsible for learning the higher order features. Max-pooling layers increase the field of view, but downsample the feature maps. Features computed by higher (deeper) convolutional layers are more abstract. However, these features lack fine details that are important for segmentation. Since the feature maps are downsampled, we need to map the lower resolution feature maps back to the input patch resolution. This is done by upsampling. As we upsample feature maps, we sum them with the feature maps of equivalent size of lower layers of the encoder path. Further convolutional layers fuse the lower and higher level features. We also employ residual blocks with pre-activations [5] that make training of deep networks easier. The last layer is a $1 \times 1 \times 1$ convolutional layer, with sofmax activation.

2.2   Glioma Grading CNN

We train a glioma grading CNN with similar architecture for each ROI (Fig. 1, middle). The ROI is extracted from each MRI sequence and resized to 96^3, before feeding it to the CNN. In these architectures, we also employ residual convolutional blocks with pre-activations [5], which contribute for better learning. After the convolutional feature computation layers, we use Global Average Pooling to summarize each feature map. Then, a cascade of $1 \times 1 \times 1$ convolutional layers act as fully-connected layers. Finally, the last layer outputs a probabilistic prediction

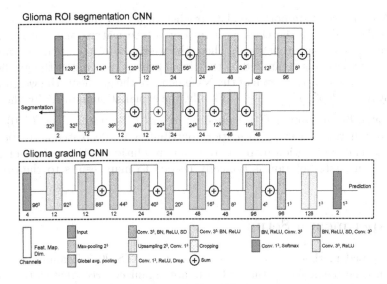

Fig. 1. Architectures of the CNNs used for glioma segmentation (top), and tumor grade classification (middle). Description of each block can be found in the bottom. BN stands for batch normalization, SD for spatial dropout [14], and Drop. for dropout.

of the tumor grade. Given the amount of available data, we use aggressive on-the-fly data augmentation during training. The data augmentation procedures were: sagittal flipping, rotation of $[-20°, 20°]$, $90°$ rotation, and exponential intensity transformation with random $\gamma \in [0.85, 1.15]$.

2.3 Grade Prediction Interpretability

To perform quality assessment of tumor grade prediction, we use the interpretability methods Guided Backpropagation (GBP) [12] and Gradient-weighted Class Activation Mapping (GradCAM) [11], after extending them to 3D. This is done at prediction time.

Guided Backpropagation [12] is based on the idea that the gradient with respect to the input image, visualized in the image space, is informative of which parts of the image are more discriminative for the neurons activation. It starts by computing a forward pass through the network layers. During backpropagation, the true gradient is not calculated. Instead, a variation that results in better explanations of ReLU activations is used. This is performed by zeroing both the gradients in the units with 0 value after ReLU activation, and the negative gradients. In this way, the backward signals of neurons that contribute for decreased activation are discarded. Although visually discriminative, GBP has the disadvantage of not being discriminative in relation to the predicted class (i.e. it can highlight areas of interest to the network but not to which class).

In contrast to GBP, GradCAM is class discriminative, but the explanation maps may have lower resolution. GradCAM tries to explain how the feature maps F of a layer l support the class prediction y^c. To that end, the gradient of the unit predicting the class with respect to the feature maps of the layer of interest $\frac{\partial y^c}{\partial F^l}$ is backpropagated. Then, the weight α_l of each feature map for the class prediction is computed as the global average pooling of the gradients. Being i, j, k the indices of each of the N elements of the gradient, the weights are given by $\alpha_l^c = \frac{1}{N}\sum_i \sum_j \sum_k \frac{\partial y^c}{\partial F^l_{ijk}}$. Finally, the explanation map E^c for the class is generated by the sum of F^l weighted by α_l^c, as $E^c = \max\left(\sum_l \alpha_l^c F^l, 0\right)$. The $\max(\cdot, 0)$ function discards information contributing for decreased activation for the class. The explanation map has the same resolution as the feature maps of interest, thus, interpolation is typically needed to map results to the original image space.

3   Experimental Setup

The proposed methods were evaluated using BRATS 2017 Training set [1,8], which has the particularity that subjects are organized according to the tumor grade into HGG (GBM) and LGG. There are 285 pre-operative acquisitions: 210 HGG, and 75 LGG. For each subject there are 4 MRI sequences available with 1 mm isotropic resolution: T1, post-contrast T1 (T1c), T2, and FLAIR. All sequences are already aligned, and skull stripped. We randomly divided the 285 subjects into 60% training, 20% validation, and 20% testing[1]. The manual segmentations of the different tumor compartments were merged into a single label to train the whole tumor segmentation network.

Two pre-processing steps are applied: bias field correction [15], and standardization of the image intensities inside the brain mask to zero mean and unit variance. All networks were trained with the Adam optimizer and crossentropy loss. For the whole tumor segmentation, learning rate (LR) was set to 5×10^{-5}, spatial dropout probability to 0.05, and weight decay to 1×10^{-6}. Regarding the CNNs for tumor grade prediction, the hyperparameters of the network were: LR -1×10^{-4}, dropout probability -0.4, and weight decay -1×10^{-4}. We used convolutional operations without padding, therefore, in skip connections, we cropped the feature maps to the same size of the smaller ones, before summing. During training, the bounding box of tumor ROI was defined using the manual segmentations. The grading CNNs were implemented with PyTorch and experiments were conducted using a NVIDIA GeForce Titan Black GPU.

For evaluation of tumor grading, we computed precision, recall, and f1-score. Since these metrics are influenced by class imbalance, we provide them for both LGG and HGG. Additionally, we compute the accuracy (acc) and the area under the receiver operating characteristic curve (ROC-AUC), which provide insights on the general ability of the classifier to distinguish between the classes.

---

[1] Grades' proportions were maintained in each set. The subjects id in each set are available online: https://github.com/sergiormpereira/brain_tumor_grading.

4   Results and Discussion

Table 1 shows quantitative results for tumor grade prediction from each of the ROI (whole brain, and tumor). We note that it is expected to achieve lower f1-score, precision, and recall for LGG, since it is the minority class. Before feeding the images to the CNNs, we standardize the image intensities with zero mean and unit variance. Common approaches in the computer vision domain compute these statistics from the whole image. However, in MRI images, the background region is usually filled with 0 intensity values after skull stripping. When we standardize the intensities in the whole image, we achieve acc. of 0.895 (whole brain) and 0.877 (tumor ROI). However, from the GBP maps (Fig. 2), we observe that the CNN considers the border of brain as discriminative, which for our data should not be a predictor of tumor grade. This is probably due to high gradients, since background has negative values, after standardization. Hence, we changed our pre-processing strategy by standardizing the image intensities inside the brain mask, only. After this approach, we observed that, mostly, the CNN does not consider the brain border as relevant for tumor grading. More interestingly, this simple change considerably boosted the metrics of tumor grade prediction from the tumor ROI (Table 1). For instance, acc. and ROC-AUC improved from 0.877 and 0.8841 to 0.9298 and 0.9841, respectively. This shows an advantage of the interpretability stage, since it allowed us to identify a systematic problem and correct it; we note that the border problem would otherwise gone unnoticed, as results were already competitive.

Focusing on the variant with the standardization in the brain mask, we observe in Table 1 that grade prediction from the tumor ROI (acc – 0.9298, ROC-AUC – 0.9841) achieves better scores than grade prediction from the whole image (acc. – 0.895, ROC-AUC – 0.8913). Despite this, we note that tumor grade prediction from the whole brain achieves an acc. of 0.895, and f1-score of 0.9286, precision of 0.9286, and recall of 0.9286 for HGG. Figure 3 shows interpretability maps for some examples. We note that GradCAM provides maps with the same resolution as the feature maps of the layer of interest. We compute GradCAM maps with the output of the third (Res3) and fourth (Res4) residual blocks (Fig. 1). Figure 3(a) shows interpretability maps for grade predictions from the whole brain. In the first row, the CNN was able to correctly grade it as HGG. From the two GradCAM maps we observe that the region of tumor was considered the most discriminative. The GBP shows focus on the ventricles, but, more interestingly, on both tumor lesions. In the second row, a HGG was mistakenly classified as LGG. The GradCAM maps are dispersed across the brain, instead of focusing in the tumor. We note that GradCAM is class discriminative, so, we show maps for LGG class. The GBP map concentrates in the ventricles. We observe that the CNN for tumor grading from the whole image focus on the ventricles frequently. We know that mass effect is a feature of HGG, and the ventricles are largely affected by it [13]. Hence, the CNN may have learned that it is a predictor of malignancy. Actually, the subventricular zone is thought to be the origin of glioma cells, and nearby brain tumors are associated with worse prognosis [7]. The focus on ventricles may explain why the example in the second

Table 1. Tumor grade results for LGG and HGG in the two ROI: whole brain, and tumor. We show results for each variant of the image intensities standardization procedure.

Region	Standardization	Grade	F1-score	Precision	Recall	Acc	ROC-AUC
Whole brain	Whole image	LGG	0.8000	0.8000	0.8000	0.8950	0.8857
		HGG	0.929	0.929	0.929		
	Brain mask	LGG	0.8000	0.8000	0.8000	0.8950	0.8913
		HGG	0.9286	0.9286	0.9286		
Tumor ROI	Whole image	LGG	0.7879	0.7222	0.8667	0.8770	0.8841
		HGG	0.9136	0.9487	0.881		
	Brain mask	LGG	0.8667	0.8667	0.8667	0.9298	0.9841
		HGG	0.9524	0.9524	0.9524		

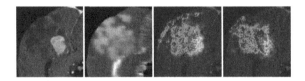

Fig. 2. Example of the effect of intensity standardization on the GBP maps. Warmer colors represent stronger responses. From left to right: T1c, T2, GBP map on image standardized over the whole image, and GBP map on image standardized in the brain region only.

row is misclassified as LGG, since its effect on ventricles is smaller than the first row example. Figure 3(b) shows examples of tumor prediction from the tumor ROI. In the first row, a HGG is correctly classified. From the GradCAM maps, we observe that the CNN correctly locates the tumor. Additionally, the Res3 and GBP maps appear to focus on the transition from necrosis to enhancing tumor and edema. This is in accordance with domain knowledge, as such an enhancing rim is characteristic for HGG. The second row of Fig. 3(b) is a LGG misclassified as HGG. In this case, it is a LGG with enhancing tumor. For this reason, the GradCAM maps for HGG and the GBP map seem to indicate that the enhancing tissues were responsible for the prediction, as it is a feature of HGG. It is possible that this is an evolving LGG that requires monitoring.

From the previous discussion, we see that GradCAM and GBP maps provide insights into the factors that contribute for a classification. So, we can see this interpretability stage as a quality assurance that enables us to check if the generated explanations are according to clinical knowledge. For instance, in the first row of Fig. 3(a) the explanations are focused on the tumor region. However, in the second row, the interpretability maps have high responses in regions that do not contain tumor. Thus, it may be a sign of an unreliable prediction, since it was based on regions of the image that are probably irrelevant. Additionally, the border effect problem, detected from the GBP maps, was a spurious pattern learned by the CNN.

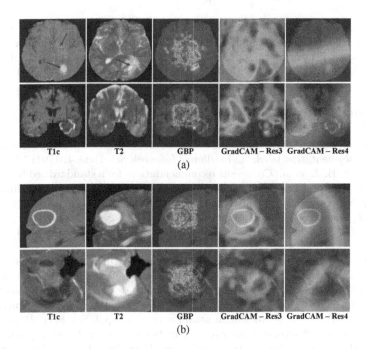

Fig. 3. Interpretability maps for grade predictions from (a) whole brain, and (b) tumor ROI. Warmer colors represent larger responses. In (a) the arrows indicate the tumor lesions; on top is a correctly classified as HGG, while example in the bottom is a HGG misclassified as LGG. In (b), the top example is a correctly classified HGG, while in the bottom a LGG is misclassified as HGG.

5   Conclusion

Tumor grading from imaging data offers a fast and non-invasive approach for anticipating tumor grading, compared with histopathological diagnosis of biopsy specimens. We propose CNN for automatic brain tumor grading from MRI images, without the need of expert ROI definition. When we predict the grade from the whole brain, we achieve acc. of 0.895, while the prediction from the tumor ROI reaches an acc. of 0.9298. Therefore, our results show that grading is possible from both ROIs, although the latter achieves substantially better scores. Additionally, we employed interpretability approaches for prediction assessment, which allowed us to improve the pre-processing stage. Moreover, it may help in assessing if a a decision is trustworthy by observing if it was actually based on the tumor region, or regions that are coherent with clinical knowledge.

Acknowledgments. Sérgio Pereira was supported by a scholarship from the Fundação para a Ciência e Tecnologia (FCT), Portugal (scholarship number PD/BD/ 105803/2014). This work is supported by FCT with the reference project UID/EEA/ 04436/2013, COMPETE 2020 with the code POCI-01-0145-FEDER-006941.

References

1. Bakas, S., et al.: Advancing the cancer genome atlas glioma MRI collections with expert segmentation labels and radiomic features. Sci. Data **4**, 170117 (2017)
2. Ellingson, B.M., et al.: Consensus recommendations for a standardized brain tumor imaging protocol in clinical trials. Neuro-Oncology **17**(9), 1188–1198 (2015)
3. Essig, M., et al.: Perfusion MRI: the five most frequently asked technical questions. Am. J. Roentgenol. **200**(1), 24–34 (2013)
4. Grier, J.T., Batchelor, T.: Low-grade gliomas in adults. Oncologist **11**(6), 681–693 (2006)
5. He, K., Zhang, X., Ren, S., Sun, J.: Identity mappings in deep residual networks. In: Leibe, B., Matas, J., Sebe, N., Welling, M. (eds.) ECCV 2016. LNCS, vol. 9908, pp. 630–645. Springer, Cham (2016). https://doi.org/10.1007/978-3-319-46493-0_38
6. Khawaldeh, S., et al.: Noninvasive grading of glioma tumor using magnetic resonance imaging with convolutional neural networks. Appl. Sci. **8**(1), 27 (2017)
7. Liu, S., et al.: Anatomical involvement of the subventricular zone predicts poor survival outcome in low-grade astrocytomas. PloS One **11**(4), e0154539 (2016)
8. Menze, B.H., et al.: The multimodal brain tumor image segmentation benchmark (BRATS). IEEE Trans. Med. Imaging **34**(10), 1993 (2015)
9. Pereira, S., et al.: Enhancing interpretability of automatically extracted machine learning features: application to a RBM-random forest system on brain lesion segmentation. Med. Image Anal. **44**, 228–244 (2018)
10. Ronneberger, O., Fischer, P., Brox, T.: U-Net: convolutional networks for biomedical image segmentation. In: Navab, N., Hornegger, J., Wells, W.M., Frangi, A.F. (eds.) MICCAI 2015. LNCS, vol. 9351, pp. 234–241. Springer, Cham (2015). https://doi.org/10.1007/978-3-319-24574-4_28
11. Selvaraju, R.R., et al.: Grad-CAM: visual explanations from deep networks via gradient-based localization. In: ICCV (2017)
12. Springenberg, J.T., et al.: Striving for simplicity: the all convolutional net. arXiv preprint arXiv:1412.6806 (2014)
13. Steed, T.C., et al.: Quantification of glioblastoma mass effect by lateral ventricle displacement. Sci. Rep. **8**(1), 2827 (2018)
14. Tompson, J., et al.: Efficient object localization using convolutional networks. In: CVPR, pp. 648–656 (2015)
15. Tustison, N.J., et al.: N4ITK: improved N3 bias correction. IEEE Trans. Med. Imaging **29**(6), 1310–1320 (2010)
16. Van Meir, E.G., et al.: Exciting new advances in neuro-oncology: the avenue to a cure for malignant glioma. CA: Cancer J. Clin. **60**(3), 166–193 (2010)
17. Zacharaki, E.I., et al.: Classification of brain tumor type and grade using MRI texture and shape in a machine learning scheme. Magn. Reson. Med. **62**(6), 1609–1618 (2009)

Visualizing Convolutional Neural Networks to Improve Decision Support for Skin Lesion Classification

Pieter Van Molle[✉], Miguel De Strooper, Tim Verbelen, Bert Vankeirsbilck, Pieter Simoens, and Bart Dhoedt

IDLab, Department of Information Technology, IMEC,
Ghent University, Ghent, Belgium
{pieter.vanmolle,miguel.destrooper,tim.verbelen,bert.vankeirsbilck,
pieter.simoens,bart.dhoedt}@ugent.be

Abstract. Because of their state-of-the-art performance in computer vision, CNNs are becoming increasingly popular in a variety of fields, including medicine. However, as neural networks are black box function approximators, it is difficult, if not impossible, for a medical expert to reason about their output. This could potentially result in the expert distrusting the network when he or she does not agree with its output. In such a case, explaining why the CNN makes a certain decision becomes valuable information. In this paper, we try to open the black box of the CNN by inspecting and visualizing the learned feature maps, in the field of dermatology. We show that, to some extent, CNNs focus on features similar to those used by dermatologists to make a diagnosis. However, more research is required for fully explaining their output.

Keywords: Deep learning · Visualization · Dermatology
Skin lesions

1 Introduction

Over the past few years, deep neural network architectures—convolutional architectures in particular—have time and again beaten state-of-the-art on large-scale image recognition tasks [6,9,14,16]. As a result, the application of convolutional neural networks (CNN) has become increasingly popular in a variety of fields. In medicine, deep learning is used as a tool to assist professionals of various subfields in their diagnoses, such as histopathology [11], oncology [1,4,17], pulmonology [7,15], etc[1]. In the subfield of dermatology, CNNs have been applied to the problem of skin lesion classification, based on dermoscopy images, where they set a new state-of-the-art benchmark, matching—or even surpassing—medical expert performance [2,3,5].

---

[1] We refer the reader to [10] for an in-depth survey on deep learning in medical analysis.

© Springer Nature Switzerland AG 2018
D. Stoyanov et al. (Eds.): MLCN 2018/DLF 2018/iMIMIC 2018, LNCS 11038, pp. 115–123, 2018.
https://doi.org/10.1007/978-3-030-02628-8_13

The challenge remains, however, to understand the reasoning behind the decisions made by these networks, since they are essentially black box function approximators. This poses a problem when a neural network outputs a diagnosis, different from the diagnosis made by the medical expert, as there is no human interpretable reasoning behind the neural networks' diagnosis. In such a case, visualizations of the network could serve as a reasoning tool to the expert.

In this paper, we train a CNN for binary classification on a skin lesion dataset, and inspect the features learned by the network, by visualizing its feature maps. In the next section, we first give an overview of the different visualization strategies for inspecting CNNs. Section 3 describes our CNN architecture and training procedure. In Sect. 4 we present and discuss the learned CNN features and we conclude the paper in Sect. 5.

2   Related Work

In [18], the authors propose a visualization technique to give some insight into the function of the intermediate feature maps of a trained CNN, by attaching a deconvolutional network to each of its convolutional layers. While a CNN maps the input from the image space to a feature space, a deconvolutional network does the opposite (mapping from a feature space back to the image space), by reversing its operations. This is done by a series of unpooling, rectifying and filtering operations. The authors use a deconvolutional network to visualize the features that result in the highest activations in a given feature map. Furthermore, they evaluate the sensitivity of a feature map, to the occlusion of a certain part of the input image, and the effect it has on the class score for the correct class.

Two other visualization techniques are presented in [13] that are based on optimization. The first technique iteratively generates a canonical image representing a class of interest. To generate this image, the authors start from a zero image and pass it through a trained CNN. Optimization is done by means of the back-propagation algorithm, by calculating the derivative of the class score, with respect to the image, while keeping the parameters of the network fixed. The second technique aims to visualize the image-specific class saliency. For a given input image and a class of interest, they calculate the derivative of the class score, with respect to the input image. The per-pixel derivatives of the input image give an estimate of the importance of these pixels regarding the class score. More specifically, the magnitude of the derivate indicates which pixels affect the class score the most when they are changed.

Concluding, typical visualization techniques either generate a single output image, in case of the feature visualization and the generation of the class representative, or function at the pixel level of the input image, in case of the region occlusion and the image-specific class saliency visualization. However, dermatologists typically scan a lesion for the presence of different individual features, such as asymmetry, border, color and structures, i.e. the so-called ABCD-score [12]. Therefore, we inspect and visualize the intermediary feature maps of the CNN on a per-image basis, aiming to provide more familiar insights to dermatologists.

3   Architecture and Training

A common approach is to use a CNN pre-trained on a large image database such as ImageNet and then fine-tune this on the target dataset [5]. The drawback is that this CNN will also contain a lot of uninformative filters (e.g. for classifying cats and dogs) for the domain at hand. Therefore we chose to train a basic CNN from scratch, but in principle our visualization approach can work for any CNN.

Our CNN consists of 4 convolutional blocks, each formed by 2 convolutional layers followed by a max pooling operation. The convolutional layers in each block have a kernel size of 3×3, and have respectively 8, 16, 32 and 64 filters. This is followed by 3 fully connected layers with 2056, 1024 and 64 hidden units. All layers have rectified linear units (ReLU) as non-linearity.

We use data from the publicly available ISIC Archive[2], to compose a training set of 12,838 dermoscopy images, spread over two classes (11,910 benign lesions, 928 malignant lesions). In a preprocessing step, the images are downscaled to a resolution of 300×300 pixels, and RGB values are normalized between 0 and 1. We augment our training set by taking random crops of 224×224 pixels, and further augment each crop by rotating (angle sampled uniformly between 0 and 2π), randomly flipping horizontally and/or vertically, adjusting brightness (factor sampled uniformly between -0.5 and 0.5), contrast (factor sampled uniformly between -0.7 and 0.7), hue (factor sampled uniformly between -0.02 and 0.02) and saturation (factor sampled uniformly between 0.7 and 1.5).

We have trained the network for 192 epochs, with mini-batches of size 96 and used the Adam algorithm [8] to update the parameters of the network, with an initial learning rate of 10^{-4} and an exponential decay rate for the first and second order momentum of respectively 0.9 and 0.999. We have evaluated the performance of the resulting CNN on a hold-out test set, comprised of 600 dermoscopy images (483 benign lesions, 117 malignant lesions), achieving an AUC score of 0.75.

4   Feature Map Visualization

For each feature map of the CNN, we created a visualization by rescaling the feature map to the input size and overlaying the activations mapped to a transparent green color (darker green = higher activation). We identify each visualization by the convolutional layer number $(0 \ldots 7)$ and filter number. Next we inspected all visualizations and tried to relate these to typical features dermatologists scan for. Especially the last two convolutional layers of the CNN (6, 7) give us some insights into which image regions grasp the attention of the CNN.

---

[2] https://isic-archive.com/.

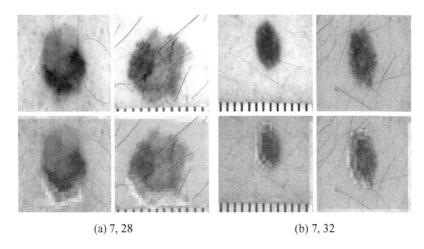

(a) 7, 28   (b) 7, 32

Fig. 1. Feature maps with high activations on lesion borders, specializing on the border location. For example, filter (a) activates on the bottom border, while filter (b) activates on the left border.

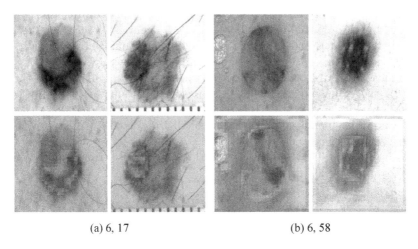

(a) 6, 17   (b) 6, 58

Fig. 2. Feature maps with high activations on darker regions within the lesion, indicating a non-uniformity in the color of the lesion.

Borders. Irregularities in the border of a skin lesion could indicate a malignant lesion. The feature maps shown in Fig. 1 both have high activations on the border of a skin lesion, but on different parts of the border. The first one (a) detects the bottom border of a lesion, while the second one (b) detects the left border.

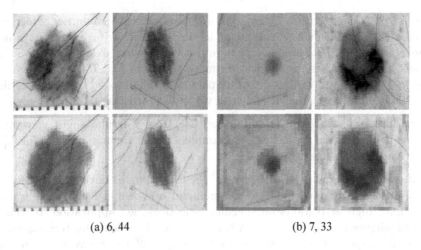

(a) 6, 44   (b) 7, 33

Fig. 3. Feature maps with high activations on skin types. For example, filter (a) activates on pale skin, while filter (b) activates on pink skin texture.

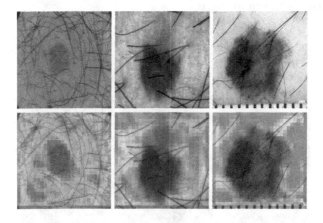

Fig. 4. A feature map (7, 8) with high activations on hair-like structures.

Color. The same reasoning tends to apply to the colors inside the lesion. A lesion that has a uniform color is usually benign, while major irregularities in color could be a sign of a malignant lesion. The feature maps shown in Fig. 2 have a high activation when a darker region is present in the lesion, implying a non-uniform color.

Skin Type. People with a lighter skin are more prone to sunburns, which can increase the development of malignant lesions on their skin. Therefore, a dermatologist takes a patient's skin type into account when examining his or her lesions. The same goes for the feature maps shown in Fig. 3. The first feature map (a) has high activations on white-pale skin. The second one (b) has high activations on a more pinkish skin with vessel-like structures.

Hair. The CNN also learns feature maps that, from a dermatologist viewpoint, have no impact on the diagnosis. For example, the feature map in Fig. 4 has high activations on hair-like structures.

Artifacts. We also noticed that some of the feature maps have high activations on various artifacts in the images. For example, as shown in Fig. 5, some feature maps have high activations on specular reflections, gel application, or rulers. This highlights some of the risks when using machine learning techniques, as this could impose a potential bias to the output of the network, when such artifacts are prominently present in the training images of a specific class.

A more elaborate overview of the activations of different feature maps on different images is shown in Fig. 6.

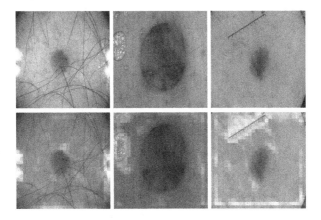

Fig. 5. Feature maps with high activations on various image artifacts. Examples are, from left to right, specular reflection, gel treatment and rulers. These artifacts could potentially impose a bias on the output of the CNN.

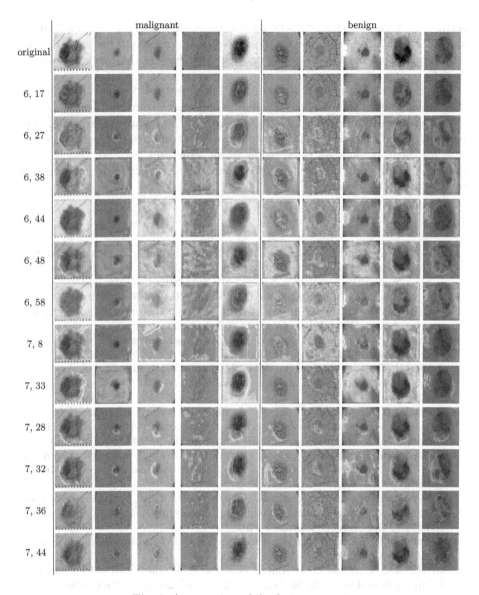

Fig. 6. An overview of the feature maps.

5   Conclusion

In this paper, we analyzed the features learned by a CNN, trained for skin lesion classification, in the field of dermatology. By visualizing the feature maps of the CNN, we see that, indeed, the high-level convolutional layers activate on similar concepts as used by dermatologists, such as lesion border, darker regions inside the lesion, surrounding skin, etc. We also found that some feature maps activate on various image artifacts, such as specular reflections, gel application, and rulers. This flags that one should be cautious when constructing a dataset for training, that such artifacts do not lead to a bias in the machine learning model.

Although this paper gives some insight in the features learned by the CNN, this does not yet explain any causal relation between the detected features of the CNN and its output. Furthermore, going through the feature maps, we did not find any that precisely highlight many of the other structures that dermatologists scan for, such as globules, dots, blood vessel structures, etc. We believe more research is required in this area in order to make CNNs a better decision support tool for dermatologists.

References

1. Cireşan, D.C., Giusti, A., Gambardella, L.M., Schmidhuber, J.: Mitosis detection in breast cancer histology images with deep neural networks. In: Mori, K., Sakuma, I., Sato, Y., Barillot, C., Navab, N. (eds.) MICCAI 2013. LNCS, vol. 8150, pp. 411–418. Springer, Heidelberg (2013). https://doi.org/10.1007/978-3-642-40763-5_51
2. Codella, N.C., et al.: Deep learning ensembles for melanoma recognition in dermoscopy images. IBM J. Res. Dev. **61**(4), 1–5 (2017)
3. Esteva, A., et al.: Dermatologist-level classification of skin cancer with deep neural networks. Nature **542**(7639), 115 (2017)
4. Fakoor, R., Ladhak, F., Nazi, A., Huber, M.: Using deep learning to enhance cancer diagnosis and classification. In: International Conference on Machine Learning, vol. 28 (2013)
5. Haenssle, H.A., et al.: Man against machine: diagnostic performance of a deep learning convolutional neural network for dermoscopic melanoma recognition in comparison to 58 dermatologists. Ann. Oncol. **29**, 1836–1842 (2018)
6. He, K., Zhang, X., Ren, S., Sun, J.: Deep residual learning for image recognition. In: IEEE Conference on Computer Vision and Pattern Recognition, pp. 770–778 (2016)
7. Hua, K.L., Hsu, C.H., Hidayati, S.C., Cheng, W.H., Chen, Y.J.: Computer-aided classification of lung nodules on computed tomography images via deep learning technique. OncoTargets Therapy **8** (2015)
8. Kingma, D.P., Ba, J.: Adam: a method for stochastic optimization. arXiv preprint arXiv:1412.6980 (2014)
9. Krizhevsky, A., Sutskever, I., Hinton, G.E.: ImageNet classification with deep convolutional neural networks. In: Advances in Neural Information Processing Systems (2012)
10. Litjens, G., et al.: A survey on deep learning in medical image analysis. Med. Image Anal. **42**, 60–88 (2017)
11. Litjens, G., et al.: Deep learning as a tool for increased accuracy and efficiency of histopathological diagnosis. Sci. Rep. **6**, 26286 (2016)
12. Nachbar, F., et al.: The ABCD rule of dermatoscopy: high prospective value in the diagnosis of doubtful melanocytic skin lesions. J. Am. Acad. Dermatol. **30**(4), 551–559 (1994)
13. Simonyan, K., Vedaldi, A., Zisserman, A.: Deep inside convolutional networks: visualising image classification models and saliency maps. arXiv preprint arXiv:1312.6034 (2013)

14. Simonyan, K., Zisserman, A.: Very deep convolutional networks for large-scale image recognition. arXiv preprint arXiv:1409.1556 (2014)
15. Sun, W., Zheng, B., Qian, W.: Computer aided lung cancer diagnosis with deep learning algorithms. In: Medical Imaging 2016: Computer-Aided Diagnosis, vol. 9785, p. 97850Z. International Society for Optics and Photonics (2016)
16. Szegedy, C., Vanhoucke, V., Ioffe, S., Shlens, J., Wojna, Z.: Rethinking the inception architecture for computer vision. In: IEEE Conference on Computer Vision and Pattern Recognition, pp. 2818–2826 (2016)
17. Wang, D., Khosla, A., Gargeya, R., Irshad, H., Beck, A.H.: Deep learning for identifying metastatic breast cancer. arXiv preprint arXiv:1606.05718 (2016)
18. Zeiler, M.D., Fergus, R.: Visualizing and understanding convolutional networks. In: Fleet, D., Pajdla, T., Schiele, B., Tuytelaars, T. (eds.) ECCV 2014. LNCS, vol. 8689, pp. 818–833. Springer, Cham (2014). https://doi.org/10.1007/978-3-319-10590-1_53

Regression Concept Vectors for Bidirectional Explanations in Histopathology

Mara Graziani[1,2(✉)], Vincent Andrearczyk[1], and Henning Müller[1,2]

[1] University of Applied Sciences Western Switzerland (HES-SO), Sierre, Switzerland
mara.graziani@hevs.ch
[2] University of Geneva (UNIGE), Geneva, Switzerland

Abstract. Explanations for deep neural network predictions in terms of domain-related concepts can be valuable in medical applications, where justifications are important for confidence in the decision-making. In this work, we propose a methodology to exploit continuous concept measures as Regression Concept Vectors (RCVs) in the activation space of a layer. The directional derivative of the decision function along the RCVs represents the network sensitivity to increasing values of a given concept measure. When applied to breast cancer grading, nuclei texture emerges as a relevant concept in the detection of tumor tissue in breast lymph node samples. We evaluate score robustness and consistency by statistical analysis.

Keywords: Interpretability · Concept vector · Histopathology

1 Introduction

Understanding representations learned by deep neural networks is a main challenge in medical imaging. Recent work on Testing with Concept Activation Vectors (TCAV) proposed directional derivatives to quantify the influence of user-defined concepts on the network output. As a real application example, the presence of diagnostic concepts such as microaneurysms and aneurysms was used to explain network predictions for diabetic retinopathy levels [6]. However, diagnostic concepts are often continuous measures that might be counter intuitive to describe by their presence or absence.

Intense research on network interpretability defined the distinction between global and local interpretability and proposed a taxonomy of desiderata, methods and evaluation criteria [1,10,12]. The relevance, or saliency, of input factors to the network decision was proposed in several gradient-based methods [12,14–16]. Outputs of these methods are typically local explanations that are gathered in attribution maps and overlayed to the original input image. The interpretability of these approaches, however, was shown to be limited and often inconsistent [7, 13]. Research in the linearity of the latent space showed that linear classifiers

© Springer Nature Switzerland AG 2018
D. Stoyanov et al. (Eds.): MLCN 2018/DLF 2018/iMIMIC 2018, LNCS 11038, pp. 124–132, 2018.
https://doi.org/10.1007/978-3-030-02628-8_14

can learn meaningful directions. These directions were mapped to semantic word embeddings in [11] or human-friendly visual concepts in [6]. TCAV computes the direction representative of a concept as the normal to the hyperplane which separates a set of concept images from a set of random images. The TCAV score estimates the influence of the user-defined concept on network decisions [6].

In this paper, we extend TCAV from a classification problem to a regression problem by computing Regression Concept Vectors (RCVs). Instead of seeking a discriminator between two concepts (or one concept and random inputs), we seek the direction of greatest increase of the measures for a single continuous concept. In particular, we compute RCVs by least squares linear regression of the concept measures for a set of inputs. We measure the relevance of a concept with bidirectional relevance scores, Br. The Br scores assume positive values when increasing values of the concept measures positively affect classification and negative in the opposite case.

We address breast cancer histopathology as an application for functionally grounded evaluation. The classification of high-resolution patches as tumorous and non-tumorous tissue is often used as a first step by state-of-the-art breast cancer classifiers [3]. Identifying the factors relevant to classification is essential to improve the physicians' trust in automated grading. For this reason, we referred to the Nottingham Histologic Grading system (NHG) [2] to select nuclear pleomorphism, and especially variations in nuclei size, shape and texture as concept measures.

The main contributions of this paper are (i) the expression of concept measures as RCVs; (ii) the development and evaluation of Br scores; (iii) the computation of nuclei pleomorphism relevance for breast cancer.

In the following, we clarify the notations adopted in the paper. We consider the set $\{\mathbf{x}_i, y_i\}_{i=1}^N$ of inputs and ground truth pairs and a deep convolutional neural network (CNN) for binary classification with prediction output $f(\mathbf{x}_i) \in [0, 1]$. The input $\mathbf{x}_i$ is a $224 \times 224 \times 3$ image patch and $y_i \in \{0, 1\}$ is the corresponding class label (with $y = 1$ for the tumor class). The disjoint set $\{\mathbf{x}_j, c_j\}_{j=1}^K$ is representative of a concept C, with measures $c_j \in \mathbb{R}$ for each image sample $\mathbf{x}_j$. In the activation space, the output of layer l for input $\mathbf{x}_i$ is $\Phi^l(\mathbf{x}_i)$ and the RCV for C is $\overrightarrow{v}_C^l$ (we will drop superscript l to simplify the notation). An overview of the method is presented in Fig. 1.

2 Methods

2.1 Correlation to Network Prediction

As a prior analysis, we compute the Pearson product-moment correlation coefficient ρ between c_j and $f(\mathbf{x}_j)$ for $j = 1, .., K$. If c_j is not relevant for $f(\mathbf{x}_j)$, their correlation should be low. In this case, $\Phi^l(\mathbf{x}_j)$ should not encode information about c_j and it should be unlikely to find a good linear regression. A high correlation could instead suggest a positive (if $\rho > 0$) or negative (if $\rho < 0$) influence of the concept on the prediction.

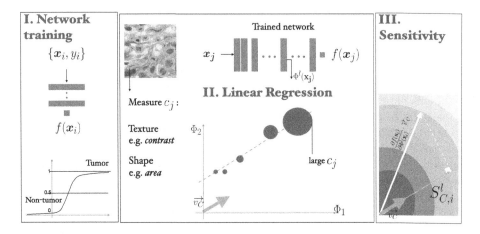

Fig. 1. *Method overview. I. Network training.* The last node of the CNN outputs a logistic regression function. The class *Tumor* is assigned to the input patch when $f(\mathbf{x}_i) > 0.5$. *II. Linear Regression.* We compute average measurements of morphological and texture features from each $\mathbf{x}_j$. Linear regression $c_j = \vec{v}_C \cdot \Phi^l(\mathbf{x}_j)$ is solved on each $(\Phi^l(\mathbf{x}_j), c_j)$ at layer l. *III. Sensitivity.* Sensitivity is computed for the $\mathbf{x}_i$ as the derivative of $f(\mathbf{x}_i)$ along $\vec{v}_C$.

2.2 Regression Concept Vectors

We extract and flatten the $\Phi^l(\mathbf{x}_j)$ for each $\mathbf{x}_j$. The RCV $\vec{v}_C$ is the vector in the space of the activation that best fits the direction of the strongest increase of the concept measures. This direction can be computed as the least squares linear regression fit of $\{\Phi^l(\mathbf{x}_j), c_j\}_{j=1}^{K}$ (see Fig. 1). In the NHG, for example, larger nuclei are assigned higher grades by pathologists. If we take *nuclei area* as a concept, we seek the vector in the activation space that points towards representations of larger nuclei.

2.3 Sensitivity to RCV

For each testing pair $(\mathbf{x}_i, y_i)$ we compute the sensitivity score $S_{C,i}^l$ as the directional derivative along the direction of the RCV:

$$S_{C,i}^l = \frac{\partial f(\mathbf{x}_i)}{\partial \Phi^l(\mathbf{x}_i)} \cdot \vec{v}_C \tag{1}$$

$S_{C,i}^l$ represents the network sensitivity to changes in the input along the direction of increasing values of the concept measures. When moving along this direction, $f(\mathbf{x}_i)$ may either increase, decrease or remain unchanged ($S_{C,i}^l=0$). The sign of $S_{C,i}^l$ represents the direction of change, while the magnitude of $S_{C,i}^l$ represents the rate of change. TCAV computes global explanations from the N sensitivities although it does not consider their magnitude. Hence, we propose Br as an alternative measure. Br scores were formulated by taking into account the principles of *explanation continuity* and *selectivity* proposed in [12]. For the former, we consider whether the sensitivity scores are similar for similar data samples. For the latter, we redistribute the final relevance to concepts with the strongest impact on the decision function. We define Br scores as the ratio between the coefficient of determination of the least squares regression, R^2, and the coefficient of variation $\hat{\sigma}/\hat{\mu}$ of the N sensitivity scores:

$$Br = R^2 \times \left(\frac{\hat{\mu}}{\hat{\sigma}} \right) \tag{2}$$

$R^2 \leq 1$ indicates how closely the RCV fits the $\{\Phi^l(\mathbf{x}_i), c_i\}_{i=1}^N$. The coefficient of variation is the standard deviation of the scores over their average, and describes their relative variation around the mean. For the same value of R^2, the Br for spread scores is lower than for scores that lay closely concentrated near their sample mean. After computing Br for multiple concepts, we scale the scores to the range $[-1, 1]$ by dividing by the maximum absolute value.

2.4 Evaluation of the Explanations

The explanations are evaluated on the basis of their statistical significance as proposed in [6]. We compute TCAV and Br scores for 30 repetitions and perform a two-tailed t-test with Bonferroni correction (with significance level $\alpha = 0.01$), as suggested in [6]. If we can reject the null hypothesis of TCAV of 0.5 for random scores and Br of 0, we accept the result as statistically significant.

3 Experiments and Results

3.1 Datasets

We trained the network on the challenging Camelyon16 and Camelyon17 datasets[1]. More than 40,000 patches at the highest resolution level were extracted from Whole Slide Images (WSIs) with ground truth annotation. To extract concepts, we used the nuclei segmentation data set in [9], for which no labels of tumorous and non-tumorous regions were available. The dataset contains WSIs of several organs with more than 21,000 annotated nuclei boundaries. From this data set, we extracted 300 training patches only from the WSIs of breast tissue.

3.2 Network Architecture and Training

A ResNet101 [5] pretrained on ImageNet was finetuned with binary cross-entropy loss for classification of tumor and non-tumor patches. For each input, the network outputs its probability to be tumor with a logistic regression function. We trained for 30 epochs with Nesterov momentum stochastic gradient descent and

---

[1] https://camelyon17.grand-challenge.org/ as of June 2018.

standard hyperparameters (initial learning rate 10^{-4}, momentum 0.9). Staining normalization and online data augmentation (random flipping, brightness, saturation and hue perturbation) were used to reduce the domain shift between the different centers. Statistics on network performance were computed from five random splits with unseen test patients.[2]

3.3   Results

Classification Performance. The validation accuracy of our classifier is just below the performance of the patch classifier used to get state-of-the-art results on the Camelyon17 challenge [3], as reported in Table 1. We report the per-patch validation accuracy for both models, although details about the training setup in [3] are unknown. Bootstrapping of the false positives was not performed and the training set size was kept limited (with 40 K patches instead of 600 K). The obtained accuracy is sufficient for a meaningful model interpretation analysis, which may be used to boost the network accuracy and generalization. Besides this, the analysis could itself be used as an alternative to bootstrapping for detecting mislabeled examples [8].

Table 1. Network accuracy % for binary classification of Camelyon17 patches.

Model	Validation accuracy
Zanjani et al.	**98.7**
ResNet101	92.43 ± 0.657

Correlation Analysis. We expressed the NHG criteria for nuclei pleomorphism as average statistics of the nuclei morphology and texture features. From the patches ($\mathbf{x}_j$) with ground truth segmentation, we computed average nuclei area, Euler coefficient and eccentricity of the ellipses that have the same second-moments as the nuclei segmented contours. We extracted three Haralick texture features inside the segmented nuclei, namely Angular Second Moment (ASM),

Table 2. Pearson correlation between the concept measurements and the network prediction.

	Correlation	ASM	Eccentricity	Euler	Area	Contrast
ρ	**−0.2985**	−0.1869	−0.1460	0.1534	0.2820	**0.4119**
p-value	≤ 0.001	≤ 0.001	≤ 0.01	≤ 0.001	≤ 0.001	≤ 0.001

---

[2] The pretrained models and the source code used for the experiments can be found at https://github.com/medgift/iMIMIC-RCVs.

contrast and correlation [4]. The Pearson correlation between the concept measurements and the relative network prediction is shown in Table 2. The concept measures for *contrast* had the largest correlation coefficient, $\rho = 0.41$.

Are We Learning the Concepts? The performance of the linear regression was used to check if the network is learning the concepts and in which layers. The determination coefficient of the regression R^2 expresses the percentage of variation that is captured by the regression. We computed R^2 for all x_j patches over multiple reruns to analyze the learning dynamics. Almost all the concepts were learned in the early layers of the network (see Fig. 2a), with *eccentricity* and *Euler* being the only two exceptions. Figure 2b shows that the concept *Euler* is highly unstable and has almost zero mean, suggesting that the learned RCVs might be random directions.

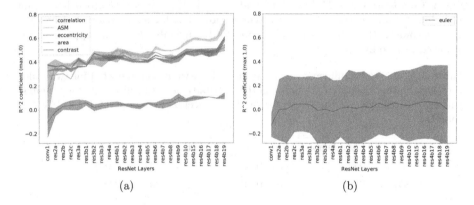

(a)   (b)

Fig. 2. (a) R^2 at different layers in the network. Results were averaged over three reruns. 95% confidence intervals are reported. (b) The RCVs for the concept *Euler* show high instability of the determination coefficient. Best on screen.

Sensitivity and Relevance. Sensitivity scores were computed on $N = 300$ patches ($\mathbf{x_i}$) from Camelyon17. The global relevance was tested with TCAV and Br, as reported in Fig. 3. *Contrast* is relevant for the classification, with TCAV = 0.75 and $Br = 0.25$. Even stronger is the impact of *correlation*, which shifts the classification output towards the non-tumor class. In this case sensitivies are mostly negative, with $Br = -1$ and TCAV = 0.1. These scores mirror the preliminary analysis of Pearson correlation in Table 2. Unstable concepts, such as *Euler* and *eccentricity*, lead to almost zero Br scores, in accordance with the initial hypothesis that the RCVs for these concepts might just be random vectors.

Statistical Evaluation. We performed a two-tailed t-test to compare the distributions of the scores against the null hypothesis of learning a random direction for the TCAV (mean 0.5) and Br (mean 0) scores. The results are presented in

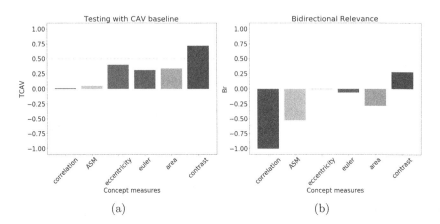

Fig. 3. Comparison of TCAV ($\in [0,1]$) and Br ($\in [-1,1]$) scores. *Contrast* is relevant according to both measurements. Br scores show that higher *correlation* drives the decision towards the non-tumor class. Scores for the unstable *Euler* are approximately flattened to zero by Br.

Table 3. Statistical significance of the scores. The p-values are reported for two-tailed t-tests evaluating the difference between the distributions of the obtained scores against a normal distribution of the scores for random concepts, i.e. mean 0.5 for TCAV and 0 for Br.

	Correlation	ASM	Eccentricity	Euler	Area	Contrast
TCAV	0.002	0.001	0.02	0.01	0.001	0.001
Br	0.001	0.001	0.30	1.0	0.001	0.001

Table 3. There was a significant difference (with $p\text{-}value \leq 0.01$) in the scores for all the relevant concepts, namely *correlation*, *ASM*, *area* and *contrast*. The statistical significance of *correlation* improves for Br scores. From the sensitivity and relevance analysis, we do not expect the *Euler* and *eccentricity* concepts to be statistically different from random directions. The analysis of both TCAV and Br scores confirms this hypothesis ($p\text{-}value \leq 0.01$) for the *eccentricity*, although the confidence to not reject the null hypothesis is higher with Br. The *Euler* concept is not rejected by the TCAV analysis. Br scores, instead, reject the hypothesis of this score being relevant.

4   Discussion and Future Work

RCVs showed that nuclei *contrast* and *correlation* were relevant to the classification of patches of breast tissue. This is in accordance with the NHG grading system, which identifies hyperchromatism as a signal of nuclear atypia. Extending the set of analyzed concepts can lead to the identification of other relevant concepts. RCVs can give insights about network training. The learning of the concepts across layers is linked to the size of the receptive field of the neurons and the increasing complexity of the sought patterns (see Fig. 2 and [6]). Hence, more abstract concepts, potentially useful in other applications, can be learned and analyzed in deep layers of the network. Moreover, outliers in the values of the sensitivity scores can identify challenging training inputs or highlight domain mismatches (e.g. differences across hospitals, staining techniques, etc.).

Overall, this paper proposed a definition of RCVs and a proof of concept on breast cancer data. RCVs could be extended to many other tasks and application domains. In the computer vision domain, RCVs could also express higher-level concepts such as materials, objects and scenes. In signal processing tasks, RCVs could be used, for instance, to determine the relevance of the occurrence of a keyword in topic modeling, or of a phoneme in automatic speech recognition.

Acknowledgements. This work was possible thanks to the project PROCESS, part of the European Union's Horizon 2020 research and innovation program (grant agreement No. 777533).

References

1. Doshi-Velez, F., Kim, B.: Towards a rigorous science of interpretable machine learning. arXiv preprint arxiv:1702.08608 (2017)
2. Elston, C.W., Ellis, I.O.: Pathological prognostic factors in breast cancer. I. The value of histological grade in breast cancer: experience from a large study with long-term follow-up. Histopathology **19**(5), 403–410 (1991)
3. Ghazvinian Zanjani, F., Zinger, S., De, P.N.: Automated detection and classification of cancer metastases in whole-slide histopathology images using deep learning (2017)
4. Haralick, R.M., Dinstein, I., Shanmugam, K.: Textural features for image classification. IEEE Trans. Syst. Man Cybern. **3**(6), 610–621 (1973)
5. He, K., Zhang, X., Ren, S., Sun, J.: Deep residual learning for image recognition. In: Proceedings of the IEEE Conference on Computer Vision and Pattern Recognition, pp. 770–778 (2016)
6. Kim, B., Gilmer, J., Viegas, F., Erlingsson, U., Wattenberg, M.: TCAV: relative concept importance testing with linear concept activation vectors. arXiv preprint arXiv:1711.11279 (2017)
7. Kindermans, P.J., et al.: The (un) reliability of saliency methods. arXiv preprint arXiv:1711.00867 (2017)
8. Koh, P.W., Liang, P.: Understanding black-box predictions via influence functions. arXiv preprint arXiv:1703.04730 (2017)
9. Kumar, N., Verma, R., Sharma, S., Bhargava, S., Vahadane, A., Sethi, A.: A dataset and a technique for generalized nuclear segmentation for computational pathology. IEEE Trans. Med. Imaging **36**(7), 1550–1560 (2017)
10. Lipton, Z.C.: The mythos of model interpretability. arXiv preprint arXiv:1606.03490 (2016)
11. Mikolov, T., Sutskever, I., Chen, K., Corrado, G.S., Dean, J.: Distributed representations of words and phrases and their compositionality. In: Advances in Neural Information Processing Systems, pp. 3111–3119 (2013)

12. Montavon, G., Samek, W., Müller, K.R.: Methods for interpreting and understanding deep neural networks. Digit. Sig. Process. (2017)
13. Ribeiro, M.T., Singh, S., Guestrin, C.: Why should I trust you?: Explaining the predictions of any classifier. In: Proceedings of the 22nd ACM SIGKDD International Conference on Knowledge Discovery and Data Mining, pp. 1135–1144. ACM (2016)
14. Selvaraju, R.R., Cogswell, M., Das, A., Vedantam, R., Parikh, D., Batra, D.: Grad-CAM: visual explanations from deep networks via gradient-based localization. In: ICCV, pp. 618–626 (2017)
15. Simonyan, K., Vedaldi, A., Zisserman, A.: Deep inside convolutional networks: visualising image classification models and saliency maps. arXiv preprint arXiv:1312.6034 (2013)
16. Zeiler, M.D., Fergus, R.: Visualizing and understanding convolutional networks. CoRR abs/1311.2 (2013). http://arxiv.org/abs/1311.2901

Towards Complementary Explanations Using Deep Neural Networks

Wilson Silva[1,2(✉)], Kelwin Fernandes[1,2], Maria J. Cardoso[2,3,4],
and Jaime S. Cardoso[1,2]

[1] Faculdade de Engenharia, Universidade do Porto, Porto, Portugal
[2] INESC TEC, Porto, Portugal
wilson.j.silva@inesctec.pt
[3] Faculdade de Ciências Médicas, Universidade NOVA de Lisboa, Lisbon, Portugal
[4] Breast Unit, Champalimaud Foundation, Lisbon, Portugal

Abstract. Interpretability is a fundamental property for the acceptance of machine learning models in highly regulated areas. Recently, deep neural networks gained the attention of the scientific community due to their high accuracy in vast classification problems. However, they are still seen as black-box models where it is hard to understand the reasons for the labels that they generate. This paper proposes a deep model with monotonic constraints that generates complementary explanations for its decisions both in terms of style and depth. Furthermore, an objective framework for the evaluation of the explanations is presented. Our method is tested on two biomedical datasets and demonstrates an improvement in relation to traditional models in terms of quality of the explanations generated.

Keywords: Interpretable machine learning · Deep neural networks
Explanations · Aesthetics evaluation · Dermoscopy

1 Introduction

In the most recent years many machine learning models are replacing or helping humans in decision-making scenarios. The recent success of deep neural networks (DNN) in the most diverse applications led to a widespread use of this technique. Nonetheless, their high accuracy is not accompanied by high interpretability. On the contrary, they remain mostly as black-box models. In this way and despite the success of DNN, in areas such as medicine and finance, which have legal and safety constraints, their use is somehow restricted. Therefore, and in order to take advantage of the DNN potential, it is critical to develop robust strategies to explain the behavior of the model. In the literature it is possible to find several different strategies to generate reasonable and perceptible explanations for machine learning model's behavior. However, those strategies can be grouped in three clusters of interpretable methods: pre-model, in-model and post-model [6].

One of the options is to consider the relevance of example-based explanations in human reasoning to try to make sense about the data we are dealing with. The

© Springer Nature Switzerland AG 2018
D. Stoyanov et al. (Eds.): MLCN 2018/DLF 2018/iMIMIC 2018, LNCS 11038, pp. 133–140, 2018.
https://doi.org/10.1007/978-3-030-02628-8_15

main idea here is that a complex data distribution might be easily interpretable considering prototypical examples. Considering that the goal is to understand the data before building any machine learning model, one can consider this strategy as interpretability before the model, i.e. pre-model.

An alternative is to build interpretability in the model itself. Inside this group, models can be based in rules, cases, sparsity and/or monotonicity. Rule-based models are characterized by a set of rules which describe the classes and define predictions. One problem typically related with this strategy is the size of the interpretable model. In order to solve this issue, Wang *et al.* [9] proposed a Bayesian framework to control the size and shape of the model. Nevertheless, a rule-based model is as interpretable as its original features are. Leveraging once more the power of examples in human understanding but now with the aim of building a machine learning model, case-based methods appear as serious competitors in the explainability challenge. In [5], the authors present a model that generates its explanations based on cluster divisions. Each cluster is characterized by a prototype and a set of defining features. From this, it is possible to deduce that the model's explanations are limited by the quality of the prototype. Sparsity is also an important property to achieve interpretability. With a limited number of activations it is easier to determine what were the events that determined the model's decision. However, if the decision can not be made with just a few activations, sparsity can decisively affect the accuracy of the model. Another way of facilitating the model interpretability is to guarantee the monotonicity of the learnt function in relation to some of the inputs [4].

Finally, interpretability can be performed after building a model. One of the options is sensitivity analysis, which consists on disturbing the input of the model and observing what happens to its output. In a computer vision context this could mean occlusions of some parts of the image [3]. One issue with sensitivity analysis is that a change in the input may not represent a realistic scenario in the data distribution. Other possibility is to create a new model capable of imitating the one which is giving the classification predictions. For instance, one can mimic a DNN with a more shallow [1] and, consequently, more interpretable network. However, it is not always the case that a simpler model exists. Lastly, we have interpretability given by investigation on hidden layers of deep convolutional neural networks [10].

1.1 Satisfying the Curiosity of Decision Makers

Human beings have different ways of thinking and learning [8]. There are people for whom a visual explanation is more easily apprehended and, on the contrary, there are people who prefer a verbal explanation. In order to satisfy all the decision makers, an interpretable model should be able to provide different styles of explanations and with different levels of granularity. Furthermore, it should present as many explanations as the decision maker needs to be confident about his/her decisions. It is also important to mention that some observations require more complex explanations than others, which reinforces the idea of different depth in the explanations.

2 Complementary Explanations Using Deep Neural Networks

In addition to their high accuracy in various classification problems, DNN have the ability to jointly integrate different strategies of interpretability, such as, the previously mentioned, case-based, monotonicity and sensitivity analysis. Thus, it is a model that presents itself at the forefront to satisfy the decision makers in their search for valuable and diverse explanations.

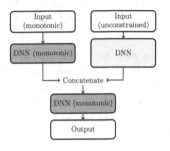

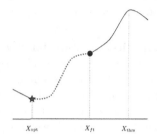

Fig. 1. Proposed DNN architecture.   **Fig. 2.** Feature impact analysis.

We will focus on binary classification settings with a known subset of monotonic features. Without loss of generality, we will assume that monotonic features increase with the probability of observing the positive class. The proposed architecture consists on two independent streams of densely connected layers that process the monotonic and non-monotonic features respectively. We impose constraints on the weights of the monotonic stream to be positive to facilitate interpretability. Then, both streams are merged and processed by a sequence of densely connected layers with positive constraints. Thus, we are promoting that the non-monotonic stream maps its feature space into a latent monotonic space. It is expected that the non-monotonic features will require additional expressiveness to transform a non-monotonic space into a monotonic one. In this sense, we validate topologies where the non-monotonic stream has at least as many –and possibly more– layers than the monotonic stream. Figure 1 illustrates the proposed architecture.

Explanation by Local Contribution. To measure the contribution, C_{ft}, of a feature ft on the prediction y, we can find the assignment X_{opt} that approximates X to an adversarial example (see (1)):

$$(\bar{y} - f(X))^2 \tag{1}$$

where $\bar{y} = 1 - y$ is the opponent class, $y \in \{0, 1\}$, and $f(X)$ is the estimated probability. We can use backpropagation with respect to ft to find the value X_{opt} (see Fig. 2) that minimizes (1). It is relevant to note that for monotonic features, such value is known a priori. Since some features may have a generalized higher contribution than others, resulting in repetitive explanations, we balanced the contribution on the target variable with the range of the feature domain traversed from the initial value to the local minimum X_{opt}. Namely:

$$C_{ft} = |f(X) - f(X')| \cdot \frac{X_{ft} - X_{opt}}{X_{\max} - X_{\min}} \quad (2)$$

where X' is the input vector after assigning X_{opt} to the feature ft. Thus, the contribution can be measured by approximating X to the adversarial space. On the other hand, the inductive rule constructed for ft covers the space between X_{ft} and the value X_{thrs} where the probability of the predicted class is maximum.

Explanation by Similar Examples. DNN are able to learn intermediate semantic representations adapted to the predictive task. Thus, we can use the nearest neighbors in the semantic space as an explanation for the decision. While the latent space is not fully interpretable, we can evaluate which features (and at which degree) impact the distance between two observations using sensitivity analysis. In this sense, two types of explanations can be extracted:

- **Similar:** the nearest neighbor in the latent space and what features make them similar.
- **Opponent:** the nearest neighbor from the opponent class in the latent space and what features make them different.

3   The Three Cs of Interpretability

Interpretability and explainability are tied concepts often used interchangeably. In this work, we will focus on local explanations of the predicted class, where individual explanations are provided for each observation. Despite the vast amount of effort that has been invested around interpretable models, the concept itself is still vaguely defined and lacks of a unified formal framework to assess it. The efficacy of an explanation depends on its ability to convince the target audience. Thus, it is surrounded by external intangible factors such as the background of the audience and its willingness to accept the explanation as a truth. While it is hard to fully assess the quality of an explanation, some proxy functions can be used to summarize the quality of a prediction under certain assumptions. Let us define an explanation as a simple model that can be applied to a local context of the data. A good explanation should maximize the following properties:

- **Completeness:** It should be susceptible of being applied in other cases where the audience can verify the validity of that explanation. e.g., the blue rows in Fig. 3 where the decision rule precondition holds and the observations within the same distance of the neighbor explanation (Fig. 3).
- **Correctness:** It should generate trust (i.e., be accurate). e.g., the label agreement between the blue rows and between the points inside the n-sphere.
- **Compactness:** It should be succinct. e.g., the number of conditions in the decision rule and the feature dimensionality of a neighbor-based explanation.

A	B	Y
0	4	1
2	3	1
4	2	0
3	6	1
3	3	1

If $A \geq 1 \wedge B \leq 5 \wedge B \geq 2$
    then $y = 1$

Compl: 3/5
Corr: 2/3
Compt: 3

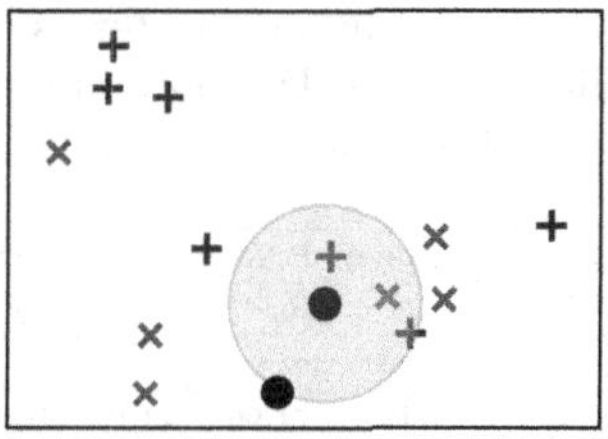

Compl: 4/14
Corr: 3/4
Compt: 2

Fig. 3. Illustration of explanation quality for decision rules and KNN (where the black dot is the new observation and the blue dot is the nearest-neighbor).

4 Experimental Assessment

We validate the performance of the proposed methodology on two applications. First, we consider the post-surgical aesthetic evaluation (i.e., poor, fair, good, and excellent) of breast cancer patients [2]. The dataset has 143 images with 23 high-level features describing breast asymmetry in terms of shape, global and local color (i.e., scars). The second application consists on the classification of dermoscopy images in three classes: common nevus, atypical nevus and melanoma. The dataset [7] has 14 features from 200 patients describing the presence of certain colors on the nevus and abnormal patterns such as asymmetry, dots, streaks, among others. In both cases, we consider binary discretizations of the problem (see Table 1). In this work, we assume features are already extracted in a previous stage of the pipeline. However, the entire pipeline covering feature extraction and model fitting could be learned end-to-end using intermediate supervision on the feature representation.

We compare the performance of the proposed DNN against classical interpretable models: a decision tree (DT) with bounded depth learned with the CART algorithm, and a Nearest Neighbor classifier (KNN with $K = 1$). We used stratified 10-fold cross-validation to choose the best hyper-parameter configuration and to generate the explanations. We explore DNN topologies with depth between 1 and 3 per block (see Fig. 1). We show in Table 1 the model performance of the three models. DNN achieved better performance than the remaining classifiers in most cases.

To measure the quality of the explanations we used accuracy for correctness, the fraction of the training set covered by the explanation as completeness, and the size in bytes of the explanation (the lower the better) after compression using the standard Deflate algorithm. Despite this compactness metric doesn't reflect the actual complexity of the explanations, it is a proxy function to define it under the assumption that the time to understand an explanation is proportional to its length. We generate explanations that account for 95% of the feature impact and embedding distance. This value can be adapted to produce more general/global or customized/local explanations. As can be seen in the

Table 1. Quality of the predictions in terms of area under the ROC and Precision-Recall curves. Quality of the explanations in terms of correctness (Corr), completeness (Compl), and compactness (Compt).

Binarization	Model	Predictions		Explanations			
		ROC	PR	Type	Corr	Compl	Compt
BCCT [2]: Breast aesthetics							
Excellent *vs.* Good, Fair, Poor	DT	71.96	92.19	Rule	75.52	3.82	31.97
	1-NN	67.37	90.74	Similar	89.27	3.25	95.94
				Opponent	72.96	80.84	96.00
	DNN	**80.61**	**96.55**	Similar	85.69	95.20	124.94
				Opponent	92.04	46.87	149.68
				Rule	99.91	3.69	62.59
Excellent, Good *vs.* Fair, Poor	DT	85.18	75.20	Rule	51.75	3.16	30.00
	1-NN	52.81	39.49	Similar	85.69	2.98	95.94
				Opponent	54.76	91.26	95.97
	DNN	**86.78**	**82.82**	Similar	72.52	17.34	80.36
				Opponent	81.16	31.28	138.00
				Rule	98.89	2.33	48.59
Excellent, Good Fair *vs.* Poor	DT	**94.20**	**74.92**	Rule	76.92	6.71	17.0769
	1-NN	54.42	20.63	Similar	94.45	3.01	95.94
				Opponent	84.42	85.33	96.00
	DNN	91.03	73.00	Similar	87.25	1.46	79.79
				Opponent	92.82	67.86	157.81
				Rule	99.88	5.48	58.44
PH^2 [7]: Dermoscopy images							
Common *vs.* Atypical, Melanoma	DT	97.60	97.90	Rule	43.00	5.03	13.10
	1-NN	94.37	94.29	Similar	94.97	5.56	15.29
				Opponent	59.42	81.38	15.94
	DNN	**99.74**	**99.83**	Similar	97.11	39.00	19.32
				Opponent	74.59	70.61	37.69
				Rule	98.86	38.83	16.27
Common, Atypical *vs.* Melanoma	DT	95.55	81.63	Rule	82.00	5.82	19.00
	1-NN	80.94	63.67	Similar	94.81	5.70	15.23
				Opponent	69.75	86.98	21.25
	DNN	**96.02**	**89.30**	Similar	91.49	8.15	33.27
				Opponent	84.02	62.12	46.24
				Rule	97.89	44.84	23.65

results, the proposed model is able to achieve the best performance in correctness results for rule explanations. For case-based explanation, the 1-NN approach with similar prototype achieves better performance in some cases at the expense of completeness. Therefore, we validate that besides having a good predictive performance in terms of classification, we can use DNN to produce explanations with high quality. Figure 4 shows some explanations produced by the DNN for both datasets.

Rule: High visibility of the scar ($sX2a > 0.98$), low inter-breast overlap ($\overline{pBOD} \leq 0.9$), low inter-breast compliance ($\overline{pBCE} \leq 0.43$) and high upward nipple retraction ($pUNR > 0.71$)

Input image

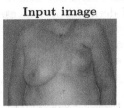

Prediction:
{Poor, Fair}

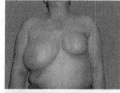
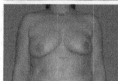
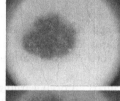

Similar case

Why?: Similar scar ($sEMDL$), inter-breast overlap ($pBOD$), color ($cEMDb$), contour difference ($pBCD$) and upward nipple retraction ($pUNR$).

Opponent case

Why?: Strong difference on the scar visibility ($sX2a$), breast overlap ($pBOD$), upward nipple retraction ($pUNR$), compliance evaluation ($pBCE$) and lower contour ($pLBC$)

Rule: It is symmetric, doesn't have black color, blue whitish veil, atypical pigmented network or streaks.

Input image

Prediction:
{Common, Atypical}

Similar case

Why?: Both images have light and dark brown color and atypical presence of dots/globules.

Opponent case

Why?: It doesn't have light brown color or atypical dots/globules. It has blue whitish veil and pigmented network.

Fig. 4. Visualization of the explanations. In the BCCT dataset we are considering the binary classification problem: {Poor, Fair} *vs.* {Good, Excellent}. Regarding the PH[2], the classification problem comes down to {Common, Atypical} *vs.* {Melanoma}. $\overline{pBOD}$ and $\overline{pBCE}$ represent the negation of the original features, $pBOD$ and $pBCE$, and are presented to make the explanation more intuitive.

5   Conclusion

In order for a machine learning model to be adopted in highly regulated areas such as medicine and finance, it needs to be interpretable. However, interpretability is a vague concept and lacks an objective framework for evaluation.

In this work, we proposed a DNN model able to generate complementary explanations both in terms of type and granularity. Moreover, there can be as many explanations as the ones the decision maker considers necessary to satisfy his/her doubts. We also define some proxy functions that summarize relevant aspects of interpretability, namely, completeness, correctness and compactness. This way we get an objective framework to evaluate the explanations generated.

The model is evaluated in two biomedical applications: post-surgical aesthetic evaluation of breast cancer patients and classification of dermoscopy images. Both the quantitative and qualitative results of our model show an improvement in the quality of the explanations generated compared to other interpretable models. Future work will focus on extending this model to ordinal and multiclass classification.

Acknowledgements. This work was partially funded by the Project "NanoSTIMA: Macro-to-Nano Human Sensing: Towards Integrated Multimodal Health Monitoring and Analytics/NORTE-01-0145-FEDER-00001" financed by the North Portugal Regional Operational Programme (NORTE 2020), under the PORTUGAL 2020 Partnership Agreement, and through the European Regional Development Fund (ERDF).

References

1. Ba, J., Caruana, R.: Do deep nets really need to be deep? In: Advances in Neural Information Processing Systems 27, pp. 2654–2662. Curran Associates, Inc. (2014)
2. Cardoso, J.S., Cardoso, M.J.: Towards an intelligent medical system for the aesthetic evaluation of breast cancer conservative treatment. Artif. Intell. Med. **40**, 115–126 (2007)
3. Fernandes, K., Cardoso, J.S., Astrup, B.: A deep learning approach for the forensic evaluation of sexual assault. Pattern Anal. Appl. **21**, 629–640 (2018)
4. Gupta, M., et al.: Monotonic calibrated interpolated look-up tables. J. Mach. Learn. Res. **17**(109), 1–47 (2016)
5. Kim, B., Rudin, C., Shah, J.A.: The Bayesian case model: a generative approach for case-based reasoning and prototype classification. In: Advances in Neural Information Processing Systems 27, pp. 1952–1960. Curran Associates, Inc. (2014)
6. Kim, B., Doshi-Velez, F.: Interpretable machine learning: the fuss, the concrete and the questions. In: ICML Tutorial on Interpretable Machine Learning (2017)
7. Mendonça, T., Ferreira, P.M., Marques, J.S., Marcal, A.R.S., Rozeira, J.: PH2 - a dermoscopic image database for research and benchmarking. In: 2013 35th Annual International Conference of the IEEE Engineering in Medicine and Biology Society (EMBC), pp. 5437–5440, July 2013
8. Pashler, H., McDaniel, M., Rohrer, D., Bjork, R.: Learning styles: concepts and evidence. Psychol. Sci. Public Interes. **9**(3), 105–119 (2008)
9. Wang, T., Rudin, C., Doshi-Velez, F., Liu, Y., Klampfl, E., MacNeille, P.: A bayesian framework for learning rule sets for interpretable classification. J. Mach. Learn. Res. **18**(70), 1–37 (2017)
10. Zeiler, M.D., Fergus, R.: Visualizing and understanding convolutional networks. In: Fleet, D., Pajdla, T., Schiele, B., Tuytelaars, T. (eds.) ECCV 2014. LNCS, vol. 8689, pp. 818–833. Springer, Cham (2014). https://doi.org/10.1007/978-3-319-10590-1_53

How Users Perceive Content-Based Image Retrieval for Identifying Skin Images

Mahya Sadeghi[1]([✉]), Parmit K. Chilana[1], and M. Stella Atkins[1,2]

[1] School of Computing Science, Simon Fraser University, Burnaby, BC, Canada
mahyas@sfu.ca
[2] School of Dermatology and Skin Science, University of British Columbia,
Vancouver, BC, Canada

Abstract. Content-Based Image Retrieval (CBIR) is an application of computer vision techniques for searching an existing database for visually similar entries to a specific query image. One application of CBIR in the dermatology domain is displaying a set of visually similar images with a pathology-confirmed diagnosis for a given query skin image. Recently, CBIR algorithms using machine learning with high accuracy rates have gained more attention since researchers have reported they have the potential to help physicians, patients, and other users make trustworthy and accurate classifications of skin diseases based on visually similar cases. However, we do not have many insights into how interactive CBIR tools are actually perceived by end users. We present the design and evaluation of a CBIR user interface and investigate users' classification accuracy on dermoscopy images and explore users' perception of confidence and trust. Our study with 16 novice users for a given set of annotated dermoscopy images indicates that, in general, CBIR enables novices to make a significantly more accurate classification on a new skin lesion image from four commonly-observed categories: Nevus, Seborrheic Keratosis, Basal Cell Carcinoma, and Malignant Melanoma.

Keywords: Dermatology · Skin cancer · CBIR · Machine learning
Artificial intelligence · Evaluation · Human-computer interaction

1 Introduction

Skin cancer is one of the most common cancers, and the number of skin related patient visits in primary care is considerable. Melanoma, the deadliest type of skin cancer, is curable if it is diagnosed early. Basal cell carcinoma, another type of skin cancer, also needs early detection to be properly treated. Considering significant number of dermatology related visits in clinics, supporting non expert physicians in their diagnostic decision can improve patient outcomes and at the same time can save costs for healthcare systems by reducing unnecessary referrals and providing early diagnosis. This can also lead to a better resource management where there is a limited access to specialists and support general

© Springer Nature Switzerland AG 2018
D. Stoyanov et al. (Eds.): MLCN 2018/DLF 2018/iMIMIC 2018, LNCS 11038, pp. 141–148, 2018.
https://doi.org/10.1007/978-3-030-02628-8_16

physicians as an educational tool. Recent advances in computer-aided diagnostic methods can aid self-examining approaches based on images, which can significantly improve early detection as the most important step to improve prognosis. In fact, modern machine learning classifiers are becoming increasingly capable of classifying skin cancer images with a level of competence comparable to dermatologists [1,2]. Although medical imaging diagnostics can benefit from intelligent computer vision and machine learning techniques, most AI algorithms provide a black box diagnosis based on percentages which clinicians do not trust [3] and most of the knowledge contained in visual data is barely extracted and applied to deliver an accurate diagnostic decision.

With recent advances in machine learning algorithms, there has been renewed interest in content-based image retrieval (CBIR) approaches where computer vision methods can be used to visually search for images to a "query" image in large databases based on the content of the image and visual clues such as shape, color, and pattern [4]. CBIR provides similar images where user can interpret the results and determine whether they are reliable. Furthermore, within the dermatology context, this technology is designed to assist with identifying and comparing skin lesions using percentage-based classifiers. CBIR-based tools can be a safe and effective implementation and integration of artificial intelligence and machine learning algorithms in clinical workflow to be validated in a low risk clinical setting. Modern CBIR systems offer powerful possibilities for lowering the overall search time and increase retrieval accuracy and are being used in a number of scientific endeavors [5,6]. Although designing and evaluating such systems in direct collaboration with users has received only limited attention, findings in a study on CT images suggest that when interpretation was supplemented with an image retrieval tool, diagnostic accuracy was improved [7]. Therefore, there are several open questions about how these tools can be safely integrated and accepted in real-world settings to support the diagnostic work of medical professionals.

In our research, we are examining how a CBIR decision support tool can be used by non-dermatologists in classification of dermoscopic skin lesion images. In this paper, we use an intuitive and scalable method on CBIR as an explainable artificial intelligence application, and investigate to what extent a CBIR system can help a non-dermatologist make an accurate classification of a given skin lesion image. We also explored to what extent the use of CBIR affects the confidence levels of these users. Our findings shed new insights into how user-centered design techniques can improve non-expert user interaction with CBIR systems and open up new opportunities for non-experts to explore, trust, and learn from medical image collections.

2   Method

2.1   Study Design

We used an experimental approach to answer our key research question: to what extent, does using a CBIR system affect user's ability to make a more accurate classification on a new skin image? The key concepts we are using to answer this question are decision accuracy, confidence level, and user trust. Our study used a within-subjects design where all the participants went through the same tasks and questionnaires. The experiment consisted of two conditions (without CBIR and with CBIR). Each user was presented with the query images one at a time and was asked to choose one best category by clicking on the appropriate button. The same normal lighting condition with a large screen was provided for all users.

2.2 Dataset

All the images are from publicly available datasets, including The International Skin Imaging Collaboration (ISIC) archive [8] and a dermoscopy atlas [9]. Since the number of skin lesion classification categories is very large (over 100 commonly observed), we had to limit our study to 4 common skin lesion categories, similar to those used in the ISIC classification challenge 2017 [10] i.e. Nevus, Seborrheic Keratosis (SK), Basal Cell Carcinoma (BCC) and Malignant Melanoma (MM). All the images were approved by an expert dermatologist who had experience working with dermoscopic images. To simplify complex medical terms for general users, we used simple terminologies for each skin lesion category. From the 1021 images in our dataset, 40 query images were chosen: 20 without CBIR and 20 with CBIR. We selected 5 query images from each category for each condition to provide an equal disease distribution. Among the 20 images in both conditions, 4 of the images were repeated, one from each category so there were 36 unique query images.

2.3 System Description

We used an existing classifier and built a user interface on top. This system was designed as a decision support tool to read and retrieve all the similar images for each query image based on a list of classification probabilities from a classifier trained on the 4 classes of interest for each image. All the images were presented to the user based on a file that stored a dictionary where the key was the name of the query image and the value is a list of tuples (imagename, cosinedistance) of the top 20 closest images inclusive. The number represented the cosine distance between that image and the query image computed using the deep feature of the query image and the image being compared. All the retrieved similar images were sorted by their cosine distance in ascending order, so the first similar image was the most similar image to the query image based on our machine learning algorithm [11]. Figure 1 shows a screen capture of the interactive user interface with our CBIR system. During the CBIR condition, the 15 most visually similar images of the collection were returned for each query image, sorted from top left row to the bottom right row. Our user-interface software for the study was written using HTML, CSS, JavaScript, NodeJS and MongoDB.

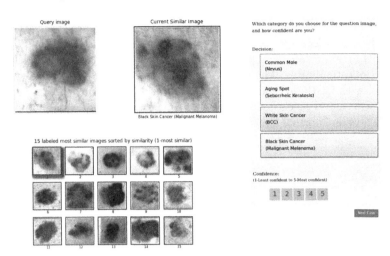

Fig. 1. Sample screenshot with CBIR algorithm results.

2.4   Protocol

We used the following protocol: After signing a consent form, participants were given a pre-task questionnaire about their past experience in medical image search. For the next 10 min, they went through a brief tutorial to learn about 4 different skin lesion categories (presented as educational slides). Next, participants started the study by classifying 20 query skin lesion images in the first condition, followed by classifying 20 query skin lesion images in the second condition. To reduce possible bias resulting from fatigue or learning effects, which are common in within-subject studies, each participant was randomized to start without CBIR or with CBIR condition. In addition, the order of "query" images was randomly selected by a shuffle algorithm inside the system, and was varied from user to user. Once the study ended, they were provided with total feedback on their performance. Finally, they filled out a post-task questionnaire about their experience.

2.5   Data Collection

We used multiple methodologies to gain insights from the different data types obtained in the study and recorded by the system. Qualitative data was obtained from interviews and questionnaire, and quantitative data such as decision accuracy and confidence level were recorded in a computer log captured in our decision support tool interface.

3   Results

16 participants successfully completed the lab experiment, including 10 males and 6 females, all non-expert adults (graduate students). From the pre-task

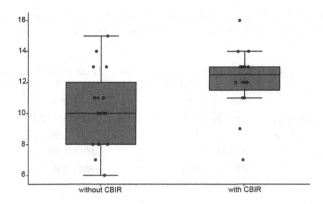

Fig. 2. Total accuracies in each condition (without CBIR and with CBIR) are shown. x axis represents the condition and y axis represents number of correct answers in each condition (out of 20). By incorporating CBIR, the mean accuracy is increased from 10.31(51.56%) to 12.19(60.94%)

questionnaire, we learned that 11 participants (68.57%) had experience in medical image search previously. They were mainly looking for photographs. The key motivation for them was personal-diagnosis (73%) and self-education (73%). However, most of the participants only found their previous searches somewhat useful and somewhat trustworthy. Irrelevant and untrustworthy images were stated as the major problems encountered during the search process.

Accuracy: For accuracy calculations, user decisions were compared to the diagnosis for every query. Overall, there was a significant improvement in mean classification accuracy from 51.56% (165 of 320) without CBIR to 60.94% (195 of 320) with CBIR as shown in Fig. 2. Corresponding null hypothesis was that there is no difference in the means, and difference in mean accuracy between conditions was tested by the two-tailed t test for paired samples. The improvement was greatest for the Nevus and MM categories, as shown in Table 1. For the Seborrheic Keratosis Category, although the accuracy decreased, no significant difference was found.

Confidence and Trust: To determine the change in users confidence in decisions without vs with CBIR, we used the Likert scale [12] score in scale of 1 (least confident) to 5 (most confident) for every query. Our null hypothesis was that there is no difference in the means. The difference in mean confidence between conditions was tested by the two-tailed t test for paired samples. The overall mean user confidence score was 3.47 without using CBIR and 3.7 with using CBIR ($P < 0.05$). Users mean confidence in TP cases was improved by 6.59% ($P < 0.05$) which shows showing similar cases is effectively increases users confidence. However when the classification result was incorrect, the impact of showing similar cases was not as significant in increasing users confidence, and was only increased by 2.52%. In addition, significant difference between confidence on correct classifications (78.16%) and incorrect classifications (69.74%)

Table 1. TP (True Positives) and Percentage of Correct classifications With and Without CBIR. Significant results where P < 0.05 (paired two-sided t test) are shown in bold.

Skin lesion category	Total correct classifications without CBIR (N = 80)	Total correct classifications with CBIR (N = 80)
Nevus	**50(62.5%)**	**72(90%)**
Seborrheic Keratosis	29(36.25%)	19(23.75%)
Basal Cell Carcinoma	49(61.25%)	57(71.25%)
Malignant Melanoma	**37(41.25%)**	**49(61.25%)**
Total	**165(51.56%)**	**195(60.94%)**

using CBIR was found (P < 0.05). Table 2 reveals mean confidence level with and without using CBIR, as well as standard deviation errors in parentheses. Trust as another critical factor was also measured on the Likert scale score in scale of 1 (least confident) to 5 (most confident) in pre-task and post-task questionnaire. Our null hypothesis was that there is no difference in the means, and the difference was tested by the two-tailed t test for paired samples. 11 of the users had previous experience with medical image search and reported a mean of 54.5% (SD = 0.98) trust on their previous findings. After the study, these users self-reported a mean of 59.29% (SD = 1.08) trust to the CBIR results; however, the difference was not significant (p = 0.65).

Table 2. Confidence level and SD of classifications With and Without CBIR. Significant results where P < 0.05 (paired two-sided t test) are shown in bold.

Classification	Average confidence without CBIR	Average confidence with CBIR
Correct	**71.57%(0.66)**	**78.16%(0.52)**
Incorrect	67.22%(0.61)	69.74%(0.48)
Total	**69.4%(0.63)**	**74%(0.54)**

4   Discussion

Although role of AI image classifiers in medicine are undeniably positive, their inner structures are often hard to comprehend and they are not usually used in the real-world settings. CBIR decision support tools can be seen as transparent applications of AI and are likely to play a growing role in the clinical practice of dermatology since this field heavily relies on the training level and expertise of medical professionals in visual inspection of skin diseases. In our user-centered design approach, we tried to tackle the problem of skin lesion classification and users' perceptions in using CBIR. Our initial results indicate that CBIR can indeed be effective for users based on the number of correct classifications they made and the increase in their confidence levels when using a CBIR interface.

According to the data collected in our study, applying CBIR models that deliver most visually similar images within the decision support tool will help users in decision making process where the final decision can be left to discretion of the user. It is noteworthy that users' accuracy scores on SK images actually decreased. Although it's not a significant difference, this may be related to the limited number of SK images in the dataset which resulted in fewer similar images from the SK category. We are currently limited to small and public datasets that often have low quality images; however, as the database for such systems grows, system accuracy is likely to increase. Our findings indicate that there should be enough representation of different disease in dataset for CBIR systems, regardless of malignancy status, when all diseases have an equal distribution balance. Other major decision making challenges for users are imperfect accuracy rates of algorithms, quality of the images(such as contrast, lighting, size), external objects in the images(such as ruler and hair), inconsistency in force and tilt while placing the dermoscopy device on the skin [13], and insufficiently magnified images.

Our findings also demonstrate that users confidence level with seeing similar images significantly increased. Hence, patient safety needs to be addressed in real clinical settings, and we need to investigate how primary physicians can adopt CBIR in clinical setting safely for better patient care outcome and more efficient workflow. Trust as another critical factor was measured in pre-task and post-task questionnaire. According to the data collected in our study, although trust is increased because of similar results, it's not a significant value. One reason may be due to the novelty of the system. Medical tools need to have long term impact, and trust can be increased overtime with personal experience, scientific evaluation, and publications. Another reason may be related to user expertise level and lack of medical knowledge.

In this study, we were limited to non-expert users as proxies for population of medical students, general physicians, and expert dermatologists. Primary care physicians have limited training in dermatology and in most cases no training in dermoscopy which is standard of diagnosis and management for skin cancer pigmented and vascular lesions. In our study we focused on novices to understand the implications of offering interactive CBIR tools to investigate their classification accuracy. According to our results, we believe this system can help users both in image interpretation and as an educational tool, since the user will be able to view pre-diagnosed similar images.

In future work, we will consider whether the results from novices transfer to other user groups. Initial informal feedback from general physicians shows their knowledge of dermoscopic skin lesions is as limited as the novices we tested, and we plan to perform user studies with experts and with physicians in future to confirm these findings. For establishing an effective interaction between a CBIR system and users, it is key to know how CBIR tools can be safely and effectively implemented, integrated, and customized for people with different levels of expertise.

References

1. Esteva, A., et al.: Dermatologist-level classification of skin cancer with deep neural networks. Nature **542**(7639), 115 (2017)
2. Han, S.S., Kim, M.S., Lim, W., Park, G.H., Park, I., Chang, S.E.: Classification of the clinical images for benign and malignant cutaneous tumors using a deep learning algorithm. J. Invest. Dermatol. **138**, 1529–1538 (2018)
3. Ribeiro, M.T., Singh, S., Guestrin, C.: Why should i trust you?: explaining the predictions of any classifier. In: Proceedings of the 22nd ACM SIGKDD International Conference on Knowledge Discovery and Data Mining, pp. 1135–1144. ACM (2016)
4. Estrela, V.V., Herrmann, A.E.: Content-based image retrieval (CBIR) in remote clinical diagnosis and healthcare. In: Encyclopedia of E-Health and Telemedicine, pp. 495–520. IGI Global (2016)
5. Dhara, A.K., Mukhopadhyay, S., Dutta, A., Garg, M., Khandelwal, N.: Content-based image retrieval system for pulmonary nodules: assisting radiologists in self-learning and diagnosis of lung cancer. J. Digit. Imaging **30**(1), 63–77 (2017)
6. Benam, A., Drew, M.S., Atkins, M.S.: A CBIR system for locating and retrieving pigment network in dermoscopy images using dermoscopy interest point detection. In: 2017 IEEE 14th International Symposium on Biomedical Imaging (ISBI 2017), pp. 122–125. IEEE, Location (2017)
7. Aisen, A.M., et al.: Automated storage and retrieval of thin-section CT images to assist diagnosis: system description and preliminary assessment. Radiology **228**(1), 265–270 (2003)
8. The International Skin Imaging Collaboration (ISIC) Archive. https://isic-archive.com/. Accessed 10 Dec 2017
9. Argenziano, G., Soyer, H.P., De Giorgi, V., Piccolo, D., Carli, P., Delfino, M.: Interactive Atlas of Dermoscopy (Book and CD-ROM). EDRA Medical Publishing & New Media, Milan (2000)
10. Codella, N.C.F., et al.: Skin lesion analysis toward melanoma detection: a challenge at the 2017 international symposium on biomedical imaging (ISBI), hosted by the international skin imaging collaboration (ISIC). In: 2018 IEEE 15th International Symposium on Biomedical Imaging (ISBI 2018), pp. 168–172. IEEE (2018)
11. Tschandl, P., Argenziano, G., Razmara, M., Yap, J.: Diagnostic accuracy of content based dermatoscopic image retrieval with deep classification features, 28 p. (2018, under review)
12. Likert, R.: A Technique for the Measurement of Attitudes. Archives of Psychology. The Science Press, New York (1932)
13. Dreiseitl, S., Binder, M., Vinterbo, S., Kittler, H.: Applying a decision support system in clinical practice: results from melanoma diagnosis. In: AMIA Annual Symposium Proceedings, p. 191. American Medical Informatics Association (2007)

Author Index

Abi Nader, Clement 3
Alves, Victor 106
Andrearczyk, Vincent 124
Antelmi, Luigi 15
Asgari Taghanaki, Saeid 87
Atkins, M. Stella 141
Ayache, Nicholas 3, 15

Balfour, Daniel R. 61
Blitz, Ari M. 79

Carass, Aaron 79
Cardoso, Jaime S. 133
Cardoso, Maria J. 133
Chilana, Parmit K. 141
Clough, James R. 61
Codella, Noel C. F. 97

Das, Arkadeep 87
De Strooper, Miguel 115
Dhoedt, Bart 115
Distergoft, Alexander 70

Eitel, Fabian 24
Ellingsen, Lotta M. 79

Feris, Rogerio 97
Fernandes, Kelwin 133

Graziani, Mara 124
Greiner, Russ 32

Halpern, Allan 97
Hamarneh, Ghassan 87
Han, Shuo 79
Haynes, John-Dylan 24
Hind, Michael 97

King, Andrew P. 61
Kügler, David 70
Kuijper, Arjan 70

Li, Xiang 79
Lin, Chung-Ching 97
Lorenzi, Marco 3, 15

Majumdar, Aabhas 52
Marsden, Paul K. 61
Mehta, Raghav 52
Meier, Raphael 106
Moelker, Adriaan 43
Mukhopadhyay, Anirban 70
Müller, Henning 124

Niessen, Wiro J. 43

Pereira, Sérgio 106
Prieto, Claudia 61
Prince, Jerry L. 79

Reader, Andrew J. 61
Reyes, Mauricio 106
Rieke, Johannes 24
Ritter, Kerstin 24
Robert, Philippe 3, 15

Sadeghi, Mahya 141
Shao, Muhan 79
Silva, Carlos A. 106
Silva, Wilson 133
Simoens, Pieter 115
Sivaswamy, Jayanthi 52
Smith, John R. 97
Sun, Yuanyuan 43

Van Molle, Pieter 115
van Walsum, Theo 43
Vankeirsbilck, Bert 115
Vega, Roberto 32
Verbelen, Tim 115

Weygandt, Martin 24

Printed in the United States
By Bookmasters